THE INVISIBLE SCORE

How AI, Data, and Algorithms Shape Decisions, Opportunities, and Everyday Life

By Subhasit Ratnam

This is a work of fiction. Names, characters, places, organizations, and incidents are either the product of the author's imagination or are used fictitiously. Any resemblance to actual persons, living or dead, events, or organizations is purely coincidental.

The information presented in this book is for educational and storytelling purposes only and should not be interpreted as professional, financial, legal, investment, or other professional advice.

This book was written by the author with the assistance of AI tools used for research support, brainstorming, editing, and design. The story, characters, themes, and creative vision remain the author's own. The cover and certain design elements were created with AI assistance and refined by the author.

First Edition

Author website: www.subhasit.com

TABLE OF CONTENTS

CHAPTER 1

THE RESULT

Maya didn't open the email right away. She had already read it in her head a dozen times. Not the exact words — but the outcome.

A quiet sense that something had already been decided somewhere she couldn't see.

The library felt different that day. Even the smallest sounds—pages turning, chairs shifting—felt careful. Results always did this. They changed the air.

She tapped the screen. The page loaded. Then the words appeared.

Application Status: Not Selected.

For a moment she wished it had just said no—for a reason she could understand. She stared at the screen longer than she expected. The screen felt brighter than everything else around her.

It wasn't just another application. It was the one.

She remembered Saturday mornings spent volunteering while her friends were still asleep. The extra projects. The essays she had rewritten more times than she could count. The recommendation letters she had been nervous to ask for.

For nearly two years the program had lived in the background of countless small decisions. Not because someone forced her. Because she believed it was worth working toward.

The one that could have led to a scholarship. The program focused on civic leadership and public policy. Most students applied because it looked good on college applications. Maya had applied for a different reason. She liked understanding how decisions affected people.

Why some opportunities opened. Why others didn't.

At the time she hadn't realized those questions would follow her far beyond the application itself.

The one her parents didn't say out loud but still believed in. Every now and then her dad would ask how the application was going. Her mom would remind her not to put too much pressure on herself. But Maya could still hear the hope in their voices.

She exhaled slowly. This wasn't just a result. It changed what came next.

"**This could have changed everything**," she said quietly.

No explanation. No reasoning. No hint of what had gone into the decision.

Across the room a chair scraped loudly. "I got in!" Jake shouted. People gathered around him instantly.

Laughter. Relief. Energy.

Maya looked back at her screen. The words hadn't changed. She didn't feel angry. Just unsettled. Because it didn't feel like someone had looked at her and said NO. It felt like something had processed her. Not someone.

And there was no way to ask why.

She closed the screen. But the decision didn't feel close.

CHAPTER 2

THE PATTERN

By lunchtime, everyone knew. Some students replayed their acceptance messages out loud, as if hearing them again made them more real.

Laughter floated across the cafeteria.

Maya wandered through the scholarship section. Rows of books stretched across the shelves. She wasn't looking for anything specific anymore.

Just answers.

Or maybe questions.

Then she sat at their usual table with Jay and Leo. She felt strangely far away from the noise.

"Well?" Jay asked.

She didn't answer. She just turned her phone toward them.

Leo read the message carefully. Once. Then again. "That's unexpected," he said.

Maya replied quietly—"Exactly."

"Jake got in," Jay added.

Leo leaned back slightly. "That might matter."

Maya looked at him.

"How?"

She still felt the weight from the library. The quiet sense that something had already made up its mind about her.

Not a person. Something else...

Leo took a moment before answering. "Think about it this way," he said.

"***If a system has seen certain patterns before—students who succeeded—it tends to look for similar patterns again***."

Maya frowned.

"That doesn't make sense," she said. "*It doesn't feel random either*."

Leo didn't respond immediately.

"So... it's comparing people?" she asked.

"In a way," Leo said. "*It's matching patterns*."

Maya looked down at her screen again. The words hadn't changed.

"*So, it's not really judging me....*"

"It doesn't know you," Leo said. "It knows patterns like you."

Maya stared at the screen again.

"Then I want to know."

Jay looked confused.

"Know what?"

"What pattern it saw."

Neither of them answered.

Maya closed the laptop.

"If a system made the decision, I want to understand how."

For the first time since the rejection, she felt something besides disappointment.

A direction.

That didn't make it easier.

Just clearer.

She flipped to a new page in her notebook and wrote -

A system doesn't know me. It knows patterns like me.

And patterns are not the same as people.

Patterns made things clearer.

But not easier.

This time… she wasn't just thinking about the result.

She was thinking about what it was based on.

CHAPTER 3

ANOTHER DECISION

In the evening, the rejection still sat with Maya as she sat down with Jay and Leo.

Jay slid his phone across the table. "Look at this."

Two loan applications were displayed side by side. At first glance, they looked almost identical.

Same categories.

Same structure.

Same type of information.

But the outcomes were different.

Approved.

Under Review.

Jay leaned back. "I mean... if the system approved it, it probably knows something."

Maya looked at him. The words from her own rejection were still fresh. "You trust it that much?"

Jay shrugged. "It's easier than questioning everything."

Maya said quietly, "Something still doesn't add up."

Leo added, "It does to the system."

He pointed to small differences. "*Credit usage here. A recent inquiry here. Slight timing differences.*"

Maya blinked. "That's minor."

"It's still a signal," Leo said. "***Systems are sensitive to patterns, even small ones.***"

Jay scrolled further, "There's another one."

Rejected.

Maya leaned closer. "Why?"

Leo studied it, "***High utilization.***"

She looked confused.

"It's like using most of a credit card limit," Leo added. "Even if you plan to pay it back, it can still look risky."

Jay nodded slowly. "She pays it back."

He hesitated. "My dad got turned down once."

Maya looked up.

"For a car," he added. "He needed it for work."

Maya felt that differently now.

Not just a decision. A delay. Something that changed what came next.

Maya sat back. "**So, the system doesn't see intent.**"

"It sees behavior," Leo said. "And assumes the pattern continues."

Maya thought about it. "That could be me. For college. For everything after."

Jay looked at her. Neither of them said anything for a moment.

Maya brought her notebook closer. She wrote:

Small signals can carry big weight

She let the thought settle. Then added: **and small signals can change outcomes.**

She looked at the notebook as if it had some answers. The same invisible logic seemed to be everywhere.

CHAPTER 4

THE SYSTEM THAT LEARNS

That evening Maya couldn't stop thinking about the rejection. Instead of closing her laptop, she opened another tab. Then another. Search after search.

How do scholarship systems work? How do applications get scored? Can algorithms make decisions?

Most of the results were confusing. One article led to another.

Eventually she found something that caught her attention.

A simple prediction model.

The next day she showed it to Leo. “Have you seen this before?”

Leo looked surprised. “Actually… yes.”

Leo pulled out a sheet of paper and drew a simple line:

Data → Patterns → Prediction → Decision

“This is the structure,” he said.

Maya studied it. “It looks simple.”

“It isn’t,” Leo said.

He opened his laptop and pulled up a small dataset.

Student performance:

Grades. Attendance. Assignments.

He ran a simple model. Predictions appeared instantly.

Maya watched quietly. "It's using the past to predict the future."

"Yes, and assuming patterns continue."

Leo glanced at the screen. "It's like predicting who might need extra help in class, but based on past patterns, not always current reality."

Maya thought about that. "What if something changes?"

Leo looked at her. "The system doesn't know yet."

That stayed with her.

Leo continued. "It's not learning from one person. It's learning from thousands of people at once."

Maya looked up. "And then applying that pattern to someone new."

Maya sat back slowly. That changed how it felt. **"It's not really seeing me."**

"It's seeing where I fit."

Leo didn't disagree.

She looked at the screen again. "This feels like the decision was already made."

Leo nodded. "In a way, it was."

Before closing the email, Maya noticed a line she had ignored the first time. **Appeals may be submitted within 60 days of the decision.**

Maya looked at the deadline again.

Maybe she would appeal.

Maybe she wouldn't.

But she had made up her mind about one thing. Before the deadline arrived, she wanted to understand the system well enough to ask better questions. Right now, she didn't even know where to start.

She stared at it for a moment. An appeal. The word felt strange. The bigger problem was that she still didn't understand what had happened.

She closed the email. But the deadline stayed at the back of her mind.

She opened the notebook and wrote:

A system learns patterns, not meaning. Patterns don't always tell the full story.

She read it once more. It wasn't just school systems. It felt like the same idea could exist somewhere else too.

She closed the notebook. For now, there seemed to be more questions than answers.

CHAPTER 5

THE FINANCIAL PICTURE

Jay's house felt different that afternoon. The usual noise from the street seemed far away. Maya sat at the kitchen table, fingers tapping lightly on the wood. Jay's father opened his laptop slowly, as if he knew this moment mattered.

A dashboard filled the screen. Clean lines. Numbers in green and red. Charts that looked simple but carried weight.

"This is what lenders actually look at," he said.

Maya leaned in closer. "It's not just **one magic number**?"

He smiled a little. **"That number you hear about, the credit score, is only a summary — the real decision comes from everything underneath it."**

He clicked through sections.

Payment history.

Amounts owed.

Length of credit history.

New credit.

Credit mix.

Maya studied each one. "It's tracking how people behave with money."

"Yes," he said. "And then trying to guess what they'll do next."

Jay leaned back in his chair. "So, if someone always paid on time before…"

"The system expects they'll probably keep doing it," Leo added from across the table.

Maya nodded slowly. "And if they missed payments?"

"It assumes the pattern might continue," Leo said.

Maya looked at the screen again. "So, it's not really judging the whole person."

"It's reading the pattern," Jay's father said quietly.

Jay frowned. "But what if there's a real reason? Like a sudden hospital bill. Or losing a job for a few months. Stuff that happens to normal people."

Maya turned to Jay's father. "Does the system see those reasons?"

He shook his head. "Not always. It sees what happened on the accounts. Not the full story behind it."

That hit Maya hard. It felt exactly like her college application.

She leaned back. "So, two completely different people could look the same to the system."

"Yes," Leo said. "If their patterns match."

Maya stayed quiet for a moment. "**That doesn't feel completely fair**."

Jay's father didn't argue. "**It's consistent. But consistency isn't the same as complete understanding**."

He paused, then continued. "There's something even bigger at play here."

Maya looked up. “What?”

“Time,” he said simply.

He pointed to the dashboard again. “These numbers don’t just show one month or one year. They show what happens when small choices repeat.”

“Small actions,” he explained. “Paying a bill on time every single month. Keeping your card balance low. Not opening too many new accounts. Those things don’t feel important at the moment.”

Jay nodded. “Like brushing your teeth. You don’t notice it until years later…”

“Exactly,” Jay’s father said. “But in money, those small habits don’t just add up. They compound.”

He opened another window and showed a simple example.

“Imagine two people. Both start at $1,000. One saves and pays for everything on time. The other misses payments and carries balances. Over ten years, the difference becomes huge because of compound interest and how the system treats their patterns.”

Maya’s eyes widened. She imagined herself in a few years—applying for student loans, maybe renting an apartment, buying her first car.

Everything suddenly felt connected.

“What if someone makes one honest mistake?” she asked softly.

Jay’s father nodded with understanding. “One mistake can follow you. It can raise your rates and make future loans costlier - exactly when you might need help the most."

But the good news is the opposite is also true.

Steady, responsible habits built early create advantages that grow stronger every year.

Maya sat quietly. She thought about her parents. The late nights. The careful budgeting. **The times they said no to things so they could say yes to her future**…

She imagined making her own choices, small choices that could quietly shape the next ten years

She flipped the page and wrote slowly:

Small habits repeated over time don't stay small. They grow. They decide more than we notice.

Systems don't just read patterns. They follow where those patterns lead, sometimes for many years.

For the first time, the invisible score didn't feel like something was happening to her. It felt like something being built, quietly, day by day, by the choices she made right now.

CHAPTER 6

PATTERNS AND SHORTCUTS

The final bell had already rung by the time Maya caught up with Leo outside. Students streamed toward buses, parking lots, and after-school activities. Jay jogged across the sidewalk and fell into step beside them.

For a minute nobody said much. The crossing signal changed. Cars moved. People waited. People followed instructions they barely noticed.

Maya adjusted her backpack. “Can a system be accurate but still unfair?” she asked.

Leo thought about it. “Depends on what you mean by accurate.”

Jay frowned. “That sounds like a politician answer.”

Leo laughed. “I’m serious.” He pointed at the screen on Maya’s phone. “It can work for most people.”

Maya looked at the rejection again. “And still fail someone.”

Leo nodded. “Exactly.”

Maya stared at the screen. “**Accuracy doesn’t guarantee fairness**,” she whispered.

Leo paused. “I used to trust these systems completely. Until one decision didn’t make sense.”

Jay looked at him. “What happened?”

Leo hesitated. “A summer program.”

Maya looked up. “You got rejected?”

Leo nodded. “I wasn’t expecting to get in. I just wanted to understand why I didn’t.”

“What did they tell you?” Jay asked.

Leo laughed softly. “Almost nothing.”

Maya immediately understood. That sounded familiar.

“So you started looking into it?” she asked.

Leo nodded. “One question turned into another. Then another.” He shrugged. “And eventually I ended up reading way more than I intended.”

Instead of pulling out her notebook right away, Maya walked quietly for a while, letting the conversation settle. Shortcuts made systems faster but sometimes less complete.

A system could be accurate and still feel unfair.

The walk home felt a little quieter after that.

CHAPTER 7

MISSING PIECES

The cafeteria was louder than usual. Students moved between tables carrying trays and half-finished conversations. Maya spotted Jay first. Leo was balancing a laptop beside a carton of milk.

"That seems risky," Maya said.

Jay dropped his tray onto the table. "If Leo spills milk on that laptop, can we finally talk about something else for a week?"

Leo looked up. "It wasn't even close."

"You say that every time," Maya laughed.

"That's because every time is somehow a different accident," Jay said.

Leo looked at the laptop, then at the milk. "Probably."

Maya sat down. Patterns helped systems move faster. But they also left things out.

"I've been thinking," she said. "What if the most important thing isn't in the data?"

Leo didn't answer immediately. "Then the system can't use it."

Maya leaned back. "That feels like part of the picture… not the whole thing."

Jay dropped into the chair beside them. "What are we solving now?"

Maya looked at him. "Let's say someone's grades drop."

Jay shrugged. "That happens."

"Why?" Maya asked.

Jay paused. "I don't know. Maybe something's going on."

Leo leaned back slightly. "Family situation. Health. Something temporary."

Maya looked between them. "But the system doesn't know that."

"It only sees the drop," Leo said.

Jay leaned back. "It reacts to the signal…"

"Not the situation," Maya finished.

She kept thinking about that as they ate.

The system could only evaluate what it could see.

If something wasn't measured, it wasn't considered.

She thought about her application again.

The forms had captured grades, activities, and achievements.

But some of the things she cared about most had never fit neatly into a box.

And what wasn't considered could still matter.

For a while, she simply sat with that thought.

CHAPTER 8

THE TEST

The next day Maya sat down before Leo even opened his laptop. The questions from the last few days were still turning in her mind.

"Let's test it again," she said.

This wasn't really about the model anymore. She was trying to understand her own rejection. Every experiment felt like another piece of a puzzle she still couldn't see clearly.

Leo nodded and pulled up the model.

Two profiles appeared. They looked the same at first glance.

"Run it," Jay said.

The results appeared.

High likelihood. Moderate likelihood.

Maya frowned. "Something doesn't add up."

Leo pointed. "Attendance dropped slightly."

Jay blinked. "That's it?"

Leo watched her think.

Maya sat back. "In real life, someone could miss a week of school."

Jay nodded. "That happens all the time."

Leo added, "But the system doesn't know why."

Maya leaned forward. "It assumes risk."

"Yes."

She looked at the screen again. "The system isn't wrong…"

Leo finished, "It's working with what it sees."

Maya thought for a moment. Then something clicked. "If a system predicts someone is risky…"

Leo nodded. "And gives them fewer opportunities…"

Jay added, "And then they struggle…"

Maya looked at both. "The system ends up being right."

Leo didn't respond immediately. "Sometimes systems reinforce their own predictions."

Maya nodded slowly. "So, they don't just predict outcomes…"

"They can influence them," Leo said.

That didn't feel like a small detail. It felt bigger. Like the system wasn't just observing outcomes anymore. It was helping create them.

Instead of flipping straight to her notebook, Maya sat with the realization for a while.

A system reacts to signals. But signals don't explain everything. Some systems don't just predict outcomes. They can influence them.

She stared at the laptop screen one more time. The model was still running. But now she saw it differently.

CHAPTER 9

THE REVIEW

A week later Maya received an update. She didn't rush to open it. She already knew what it would say.

She tapped the screen. **Status: No Change.**

Jay leaned over her shoulder. "That's it?"

Maya nodded in silence.

Leo read it carefully. "They reviewed it."

"And nothing changed," Jay added.

Maya leaned back. "So, review doesn't mean rethinking everything."

Leo shook his head. "It often means checking the same inputs again."

Maya looked at the screen. "If nothing new is added…"

"The outcome stays the same," Leo said.

Jay frowned. "That doesn't feel like a real review."

Maya shook her head. "They checked the same inputs. That's not the same as rethinking it."

Leo replied, "That distinction matters."

A few weeks had already passed. The appeal deadline was getting closer. Jay noticed Maya staring at the scholarship page again.

"You still thinking about it?"

Maya nodded. "I don't even know what I'd write."

Jay looked confused. "An appeal?"

Maya closed the laptop. "**How do you challenge a decision when you still don't understand it?**"

Instead of writing a summary, Maya simply sat with the frustration a little longer.

The decision still stood.

Nothing had changed.

For a moment she stared at the screen.

A few weeks earlier, she had imagined a review as someone taking a fresh look. Someone noticing what had been missed.

Now she understood it differently.

If the same information went back into the system, the same answer was likely to come back out.

The result still felt disappointing.

But at least it made sense.

She closed the laptop slowly.

The review had answered one question.

The decision hadn't changed.

But now she understood why.

And somehow that felt different.

CHAPTER 10

THE GLITCH THAT ISN'T A GLITCH

A few days later, Maya sat across from Leo in the library.

The appeal deadline was still weeks away, but she wasn't thinking about writing an appeal anymore.

She was thinking about systems.

About reviews.

About decisions that didn't change.

And about the uncomfortable realization that understanding something didn't always make it feel fair.

The review had explained the decision.

But it hadn't changed it…

Leo opened his laptop.

"I want to show you something," he said.

Jay dropped into the chair beside them.

"If this is another spreadsheet, I'm leaving."

"It isn't," Leo said.

"Good. My brain is still recovering from the last one."

Maya laughed.

For the first time in days, the scholarship rejection wasn't the only thing on her mind.

Leo turned the screen toward them.

A simple prediction model appeared.

"Same one?" Maya asked.

"Mostly."

Leo entered a set of information and clicked Run.

A score appeared.

Then he changed one value.

The score barely moved.

He changed another.

Still almost no difference.

Jay shrugged.

"Seems normal."

Leo nodded.

Then he adjusted a different field.

The score dropped sharply.

Jay leaned forward.

"Wait."

Leo ran it again.

The same result appeared.

Jay frowned.

"That has to be a glitch."

Leo changed the value back.

The score returned to its original level.

He changed it again.

The score dropped.

Exactly the same way.

Maya stared at the screen.

The result wasn't changing.

It was repeating.

"If it's consistent, then maybe it's right," Jay said.

Maya shook her head.

"Or maybe it's consistently missing something."

Leo looked at the result again.

"It's consistent."

Jay frowned.

"That doesn't make it right."

"No," Maya said quietly.

"But it means it's not random."

The screen hadn't malfunctioned.

Nothing was broken.

The system was doing exactly what it had been designed to do.

She looked at the score again.

"This wasn't an accident."

She paused.

"This was always going to happen."

Leo nodded.

Jay leaned back.

"So it's built this way."

The words hung in the air.

Maya sat still for a moment.

That realization felt different.

A glitch could be fixed.

A bug could be corrected.

But this wasn't a mistake — this was part of the design.

The model wasn't accidentally producing the outcome.

It produced the outcome because of the rules it had been given.

Because of the patterns it had learned.

Because of the assumptions built into it.

She thought about the review of her application.

The outcome hadn't changed.

Not because the system had failed.

Because the system had worked exactly as intended.

That felt harder to accept.

She opened her notebook and wrote:

Not everything that feels wrong is an error.

Some outcomes come from how systems are built.

Sometimes the system works exactly as designed.

She stared at the words.

Understanding that didn't change the result.

But it changed how she saw it.

For a long time she had imagined systems as machines that occasionally failed.

Now she was beginning to understand something more complicated.

Sometimes the most important questions weren't:

"Did the system fail?"

Sometimes they were:

"Who designed it this way?"

And:

"What assumptions did they build into it?"

She closed the notebook.

The library was quiet again.

Students studied.

Pages turned.

Laptops clicked.

Nothing looked different.

Yet Maya felt as though she had noticed something she couldn't unsee.

She took a slow breath.

She wasn't angry at the outcome anymore.

She was thinking about the system itself.

And for the first time, that felt like the real question.

CHAPTER 11

THE AI THAT TALKS BACK

Leo showed them something new. “Try this,” he said.

Maya typed a question.

The response appeared instantly.

Clear. Confident.

“That sounds right,” Jay said.

Maya read it again.

“It does.”

Leo looked closely.

“That’s the point.”

Maya looked at him.

“What do you mean?”

“It’s designed to sound right,” Leo said.

Maya tried another question.

Same confidence. Different questions.

“***It never sounds unsure***,” Jay said.

Leo nodded slightly.

"When something sounds confident or official, people tend to question it less…"

He added, "That's called ***automation bias***."

Maya responded like she was talking to herself.

"Even when they should."

Leo didn't disagree.

She wrote:

The more confident a system sounds,

The less people question it.

Confidence is not correctness.

The AI sounded sure. But she was learning to question it anyway.

Questioning… even though the answers weren't in sight yet. But it felt like a step forward.

CHAPTER 12

THE INVISIBLE AUTHOR

That evening, Maya opened the tool again. She didn't hesitate this time.

"Try something real," Jay said.

Maya entered, "Write a short paragraph about why students should apply for scholarships."

The response appeared instantly.

Clear. Structured. Persuasive.

Jay read it. "That's really good."

Maya read it again. It *was* good. Better than what she had in mind.

"Try your version," Leo said.

She hesitated. Then she started typing on her own. She wrote a few lines.

Stopped.

Rewrote a sentence.

Deleted a word.

When she finished, she compared the two.

The system's version sounded smoother.

More polished.

More... complete.

Jay leaned over.

“I’d use that,” he said, pointing at the AI version.

Maya didn’t respond.

Something didn’t sit right.

“It sounds better,” she said slowly.

Leo gave a small nod

“It’s designed to.”

She looked again.

“So, what am I doing then?” she asked quietly. “Learning... or just selecting what sounds best?”

No one answered right away.

She tried something else.

She copied the AI version. Then changed a few words. Adjusted a sentence. Added one line of her own.

It still felt like the same answer.

Just slightly edited.

Jay smiled.

“That’s efficient.”

Maya leaned back.

“It is,” she said.

Then she hesitated.

"But it doesn't feel like mine."

Maya looked around the room.

Two students nearby were working on laptops.

One had an AI tool open.

"Summarize this," he typed.

The answer appeared instantly.

He copied it.

Changed a few words.

Submitted it.

The other student was doing the same thing.

Different assignments. Same process.

Jay glanced over.

"That's normal now," he said.

Maya watched for a moment.

No hesitation.

No second thought.

Just faster.

"Do they even read it?" she asked quietly.

Jay shrugged.

"Probably enough to submit."

Maya didn't respond.

It wasn't wrong.

It just felt... Incomplete.

She looked at the screen again.

If something sounds better than what you can write...

Does that make it better—or just easier?

Leo watched her carefully.

"That's the trade-off."

Maya looked at him.

"What do you mean?"

Leo spoke calmly.

"When something feels easy and confident," Leo said, "People stop questioning it."

Maya nodded. "So, **the easier it gets, the less we question it**."

Leo said, "That's the same idea-automation bias again."

She looked back at the screen again.

For the first time, the answer didn't just look helpful.

It felt... familiar.

Like something she could use without thinking too much.

That was the problem.

Maya closed the tab.

Jay frowned.

"You're not going to use it?"

Maya shook her head.

"Not like that."

She opened her bag and looked for the notebook.

She didn’t rush this time.

She wrote:

Prediction is not understanding.

That hit differently, then she added:

And something that sounds right isn’t always something we’ve learned.

She looked at the page for a moment.

Then kept the notebook back into the bag.

Now, she wasn’t just thinking about what the system could do.

She was thinking about what it might replace…

She looked around the room.

Nothing had changed.

The same desks.

The same students.

The same conversations.

And yet she couldn't stop wondering **how many invisible systems had already shaped the day.**

She didn't write the thought down.

She carried it home instead.

CHAPTER 13

MAYA'S MODEL

The next day Maya reached the library early. The room was almost empty. Only the soft hum of the air conditioning and the occasional turn of a page broke the silence.

She sat down, opened her laptop, and looked at Leo. "I want to build something myself today."

Jay dropped his bag with a grin. "This should be good."

Maya took a slow breath and began creating fields - grades, attendance, activities, hours spent studying, even a small box for personal notes.

As she typed, she kept thinking about her own application and how neatly everything had been summarized.

"This is basically what real systems use, right?" she asked.

Leo nodded. "Very close."

But the blank rows still didn't feel enough.

That evening at home, Maya did something she had been avoiding. She opened her rejected application again. Everything looked so clean. Too clean...

Two full years of effort, late nights, and quiet pressure had been summarized and simplified.

Student council meetings.

Community volunteer projects.

Weekend events she had helped organize.

Hours spent working on community projects she hoped would make a difference someday.

The application captured what she had done.

But not why she cared about any of it.

Grades listed. Activities reduced to a few short lines. Achievements compressed into checkboxes and categories.

Everything looked measurable. But something felt missing.

Her chest tightened. She remembered the weekends volunteering while her friends slept in, the essays she rewrote endlessly, and the questions she kept asking long after others had moved on.

None of that had fit neatly into the form.

Jay texted her. “Still thinking about it?”

“Yeah,” she replied. “Looking at my application again.”

“It looked fine when I saw it,” he wrote back.

“That’s exactly the problem,” Maya typed. “It looks fine on the surface. But it’s missing so much.”

The next afternoon she returned to her model. She filled in sample data and clicked run.

The prediction score appeared almost instantly.

Jay leaned in. “Whoa. That was quick.”

Maya didn't reply at first. She lowered the attendance number slightly. The score dropped more than she expected. She tried adding a strong activity. The result shifted again.

"So this isn't just about the data," she said quietly. "It's about what someone decided the data means."

Leo leaned back. "Exactly."

Jay looked confused. "So, the system isn't just learning everything by itself?"

"Not completely," Leo explained. "People choose what to measure. They decide how much weight each piece gets. Those hidden choices shape every prediction."

Maya kept experimenting. She changed several values at once. The score swung dramatically. She sat back and stared at the screen. Someone had decided attendance mattered more than activities. Someone had decided exactly how much more.

"And all of this learns from past examples, right?" she asked. "What if those examples were unfair?"

"Then the model can learn and repeat that unfairness," Leo said. "The system doesn't understand right or wrong. It just sees what happened before."

The library felt quieter. Maya added imaginary personal challenges and ran the model again. The prediction changed. She tried removing them. It changed once more.

"So, it's like something goes in," she said slowly, "and something comes out. But we don't fully see what happens in the middle."

Leo nodded. "That's what people call a **black box**."

Maya repeated the words softly. "A black box."

She stared at her simple model. The same inputs. The same basic idea. But now it felt much less trustworthy.

"Can the system ever explain exactly why it gave this answer?" she asked.

Leo shook his head. "Not always. And not in a simple way that really captures everything."

Jay looked surprised. "You can build something like this and still not fully understand how it makes decisions?"

"It happens more often than most people realize," Leo said.

Maya closed her laptop. She didn't need to run it again.

Instead, she opened her notebook and wrote one line that felt more honest than any summary:

I can see the numbers go in and the score come out, but never what happens in between. If the past was unfair, does the future have to be too?

She closed the notebook.

The model hadn't answered her original question.

But it had changed the way she looked at answers.

For the first time, she understood how much trust people placed in things they couldn't fully see - and how dangerous that trust could become when they stopped asking questions

CHAPTER 14

CONNECTING THE DOTS

Maya spread her notes across the table.

Admissions

Loans

AI tools

"It's all the same idea," she said.

Jay looked up. "How?"

"They all use patterns," Maya said.

Leo gave a quiet nod. "And the same limitations."

Maya looked at her earlier notes again. This wasn't just about systems anymore. It was about decisions that had already shaped her.

"They don't understand context." "They rely on data." "They use shortcuts."

Jay added, "And they assume patterns continue."

Maya nodded. "That explains a lot."

She took a moment. It made sense. But it didn't feel complete. The rejection, the loan applications, the confident AI answers, they all felt connected now. Like different doors leading into the same room.

She noted:

All these systems follow patterns. But patterns are not the full story.

She read it once more.

Now, she saw the connections clearly. The same invisible forces were at work everywhere. And now she was starting to notice them.

Maya sat with Jay and Leo at their usual table. The cafeteria noise felt louder than usual.

Jay took a big bite of his sandwich and grinned. "So, Maya, you're basically a detective now. Solving the mystery of the evil algorithm."

Maya smiled, but it came a second too late.

Leo stayed quiet, poking at his fries.

Jay noticed. "What? Was that not funny?"

"It was fine," Maya said quickly.

The table went quiet for a few seconds. Forks scraped against trays. Someone laughed loudly at the next table.

Jay shifted in his seat. "You've been kind of quiet lately. Everything okay?"

Maya looked down at her food. "Yeah. Just… thinking about a lot."

Leo finally spoke. "We don't have to talk about systems all the time, you know."

Jay nodded. "We can just be normal sometimes."

Maya felt a small knot in her stomach loosen. She stole one of Jay's fries.

“Sorry,” she said. “I guess I’ve been in my head.”

Jay shrugged and pushed the tray toward her. “Take as many as you want. That’s what friends are for.”

They laughed. The awkwardness melted away. For a few minutes they were just three kids eating lunch, teasing each other like always.

CHAPTER 15

LINES ON A MAP

Later that week Maya was researching lending decisions for a project. Most of what she found looked ordinary. Then she clicked on a map.

At first she didn't understand what she was looking at. Certain neighborhoods were outlined in different colors. The patterns felt strange.

The next day she brought it to the library. "Have either of you seen this before?"

Leo looked at the screen. His expression changed immediately. "Where did you find that?"

Maya didn't expect a map to matter so much. Leo pulled it up quietly on his laptop. Different neighborhoods. Different colors.

Maya leaned closer. "What is this?"

"Old lending patterns," Leo said.

Jay frowned. "You mean where people got loans?"

Leo nodded. "And where they didn't."

Maya stared at the screen. "It looks… divided."

"It was," Leo said. He zoomed in. Some areas were marked differently. "Years ago people used maps like this to decide who was 'safe' to lend to."

Maya frowned. “Based on what?”

“Location,” Leo said.

Jay blinked. “Not the person?”

Leo shook his head. “Not always.”

Maya leaned back. “That doesn’t seem fair.”

“It wasn’t,” Leo said.

The room stayed quiet for a moment. Maya stared at the map. “How do you even know this stuff?”

Leo looked slightly embarrassed. “I’ve been reading about it for a couple of years.”

“A couple of years?” Jay asked.

Leo nodded. “Once I started learning how systems make decisions, I kept running into stories about where the data came from.”

Maya looked back at the map. “And this was one of them?”

“Yeah.”

Then Maya looked at the screen again. “But systems today don’t do that… right?”

Leo didn’t answer right away. “They use more data now. More signals. More patterns.”

Maya nodded slowly. “So it’s better.”

“Sometimes,” Leo said.

Maya caught the pause. “What do you mean ‘sometimes’?”

Leo turned the laptop toward her. “Let’s think about how these systems decide.”

Maya leaned forward. "They predict risk."

Leo nodded. "Right. But predictions aren't always correct."

Jay tilted his head. "You mean they can make mistakes?"

"Two kinds happen a lot," Leo said.

Maya looked interested. Leo continued. "Imagine someone applies for a loan. They're responsible. They would repay it. But the system predicts they won't."

Jay frowned. "So they get rejected… even though they're good?"

Leo nodded. "That's called a **false negative**."

Maya repeated it quietly. "False… negative."

Leo explained. **"The system says 'no' when the correct answer should have been 'yes.' It's a missed opportunity."**

Jay leaned forward. "What's the other kind?"

Leo continued. **"Now imagine the opposite. Someone isn't likely to repay it. But the system approves them anyway."**

Jay blinked. "So it says 'yes'… when it should be 'no'?"

Leo gave a small nod. "That's a **false positive**."

The screen felt heavier now.

"So both mistakes are possible," Maya said.

"Yes," Leo said. "One blocks someone who deserved a chance. The other gives a chance where it might not work."

Maya thought about that for a moment. "So even if something feels wrong…"

"It doesn't always get corrected," she finished quietly.

Leo nodded. "That's why many important systems still keep a **human in the loop** - a real person who reviews the decision before it becomes final. The system can suggest or flag things, but a human reviews the decision before it becomes final. Especially when the stakes are high."

Jay frowned. "Does that actually help?"

"Sometimes," Leo said. "It's not perfect. But it adds judgment that data alone can't provide."

Maya sat back slowly. **Fairness isn't automatic. It depends on the data, the design, and the people responsible for the decisions.**

She couldn't ignore it now. She wasn't just asking if the system was right. She was asking who decided how it worked.

Maya looked back at the map. A few colored lines. A few old decisions. And somehow they had reached all the way into the future.

For the first time, she understood that patterns didn't appear on their own. People created them. Then systems learned from them.

She thought about the scholarship again. The application. The scores. The decisions.

Maybe the hardest question wasn't whether a system was right. Maybe it was understanding how it could be wrong. And who might be affected when it was.

CHAPTER 16

WHEN SYSTEMS FEEL UNFAIR

Maya didn't say it out loud at first. But the thought stayed with her, quiet and stubborn. *This doesn't feel fair.*

It followed her through the rest of the week. It sat with her at dinner. It lingered while she tried to study. And it came with her back to the library, where the usual quiet felt heavier than usual.

Jay was the one who finally said it. "So what happens when a system gets it wrong?"

Leo looked up from his laptop. "It depends."

Maya leaned forward, resting her chin on her hand. "On what?"

"Whether there's a way to question it."

The room grew quieter.

Jay frowned. "You mean like an appeal?"

"Sometimes," Leo said. "Sometimes there's an appeal. Sometimes a review. Sometimes another person actually looks at the decision."

"And sometimes?" Maya asked.

Leo hesitated. "Sometimes the decision just stands."

Nobody spoke for a moment. Maya felt the words settle in her chest. She thought about her scholarship application again - the waiting, the hoping, the email she had checked too many times. The review had happened. The result hadn't changed. She understood why now. But understanding didn't make it hurt any less.

She looked at Leo. "Why do some decisions feel unfair even when the system follows the rules?"

Leo leaned back slowly. "Because **rules and fairness aren't always the same thing**."

Jay blinked. "That sounds complicated."

"It is," Leo said softly.

He glanced at Maya's notebook. "Imagine a teacher grades every test exactly the same way."

"Okay."

"The rules are consistent."

Jay nodded. "That sounds fair."

"Now imagine one question is worded badly and half the class misunderstands it."

Jay paused. "Oh."

"The grading is still consistent," Leo said. "But the students might not think the result is fair."

Maya felt a small ache in her throat. She thought about all the Saturdays she had spent volunteering, the essays she had rewritten until her eyes burned, the quiet hope her parents tried not to show. The system had followed its rules. But it still felt like something important had been missed.

"Fairness isn't automatic," she said quietly.

Leo nodded. "It depends on how systems are built… and who gets to question them."

The library fell quiet again. Students studied nearby. Pages turned. A keyboard clicked steadily in the background. Everything looked normal. But Maya's mind kept turning the same thought over and over.

Who decides what is fair?

She opened her notebook and wrote slowly:

Fairness is not just about the decision. It's also about whether people can understand it, question it, and improve it.

She stared at the words for a long moment, then closed the notebook.

The question still didn't have a simple answer. But now she knew something important. A system wasn't fair just because it worked. Sometimes fairness depended on whether people were allowed to challenge it.

And right now, that felt like the only thing that mattered.

CHAPTER 17

THE PERFECT TRAP

The email arrived late. Late enough to feel important.

Subject line: **Congratulations - Ivy Scholars Program Selection**

Maya opened it immediately.

It looked real. Perfectly timed.

Jay leaned over.

"No way."

Maya didn't smile yet. She kept reading.

"You have been selected for the next stage of review," it said.

"Please confirm your details to proceed."

Her attention sharpened.

Another line appeared below.

Action required within 12 hours to retain eligibility.

She hesitated.

Jay leaned closer.

"Twelve hours?"

Maya read it again.

“Failure to respond may result in your application being withdrawn.”

The excitement shifted. **Now it felt urgent. Real.**

Her hand moved toward the link. Her finger hovered. Just for a second.

“Wait,” Leo said.

Maya stopped.

“It looks real,” she said.

For a few minutes, Maya felt embarrassed for almost falling for it.

Then something else bothered her.

Part of her still wanted the message to be real.

That was the part she trusted even less.

Leo nodded.

“That’s the point.”

A notification appeared.

Reminder: 11 hours 32 minutes remaining

Jay blinked.

“It’s counting down?”

Maya felt it now. Not just excitement, but pressure.

“It feels like I might lose something,” she said.

Leo didn’t disagree.

“**Urgency changes decisions**.”

Maya hovered over the link again. Everything about it felt right. the name, the tone, the timing. It didn't look fake. It looked important.

Another notification.

Reminder: 10 hours 58 minutes remaining

Maya leaned back slightly.

"This isn't just information. It's pressure."

Leo nodded.

"And **pressure makes people act faster than they think**."

Maya moved the cursor slowly across the link. The address appeared at the bottom. It looked right.

Almost.

But not exactly. One letter was different.

She stared at it.

"That's easy to miss," she said.

Leo nodded. "It's meant to be."

The countdown kept running. **09:47:12**

For a moment, she imagined clicking it. Typing into her details. Submitting quickly. Just to be safe. Just in case.

Jay broke the silence.

"I probably would've done it," he said.

Maya didn't look at him.

"I almost did," she replied.

"I saw something weird yesterday," Jay said after a moment, scrolling through his phone.

"What?" Maya asked.

"Someone got a message that sounded exactly like their friend asking for money."

Maya frowned.

"Like a scam?"

"Yeah, but it didn't feel like one," he said.

"It sounded real."

Leo nodded slowly.

"That's the shift. It's not just obvious scams anymore. It's messages that feel real enough to trust."

Maya paused.

"So, it's not just about making decisions," she said quietly.

"Sometimes it's about trusting things… that aren't even real."

That felt honest.

Later that evening, Jay walked in again.

"Something weird happened," he said.

Maya looked up.

"That message I got earlier..." he continued.

"I logged in."

Jay exhaled.

"I didn't even think twice."

Maya looked at him.

"That's what it wanted."

Jay nodded slowly.

"Yeah... it worked."

The room went quiet.

"And?" Maya asked.

Jay hesitated.

"Nothing happened. But then I got a password reset alert."

Maya felt the shift.

"So, they didn't break in," she said slowly.

"They didn't break in," Leo replied.

"They were invited in."

Jay nodded.

"I had to change everything."

Maya looked back at her screen. The Ivy message was still there. Still convincing. Still urgent. Still counting down.

But now it felt different.

Not exciting. Not important.

Just... designed.

She felt like writing it down…

She wrote

Not all attacks break systems. Some use trust.

She read it again.

Then added

Urgency makes something feel real before we have time to question it.

And something she learned today:

If something feels too important to wait, that's when we should slow down.

It stood out now. She wasn't just thinking about the message.

She was thinking about how easily it could have worked.

CHAPTER 18

WHEN ANSWERS SOUND RIGHT

The next day, Maya didn't go back to the model. She opened the AI tool instead.

"Ask it something tricky," Jay said.

Maya typed a question. The answer appeared instantly.

Clear. Structured. Confident.

"That sounds right," Jay said.

Maya read it again. Then tried another question. Same confidence.

"It never sounds unsure," Jay said.

Leo nodded slightly. "That's the point. It's designed to sound right."

He paused for a moment. "That's called automation bias."

Maya leaned back. "Even when we should question it."

Leo didn't disagree.

Maya opened her notebook. The pages were getting full.

The more confident a system sounds, the less people question it. Confidence is not correctness.

She stared at the words for a moment. The AI sounded certain. But certainty suddenly felt different to her now.

Later that evening, Maya looked up briefly from her phone. In the next room, she could hear her parents talking quietly about something they had seen online. One article said one thing. Another said the opposite. Even adults sounded uncertain now.

That surprised her. Growing up, parents had always seemed like they already understood how the world worked. But lately it felt like everyone was trying to figure things out at the same time.

Online, people her age moved through trends, apps, and information almost instantly. Adults moved slower. More carefully. Sometimes Maya thought that meant they understood less. But now she wasn't so sure. **Maybe they were just more used to stopping before believing something.**

"You've been on your phone a lot lately," her mom said gently from the kitchen.

Maya sighed softly. "Everybody says that."

"I'm not trying to start an argument," her mom replied. "I just worry sometimes."

Maya looked down at the screen. "You didn't grow up with this stuff. It's different now."

Her mom nodded. "I know."

The answer surprised Maya. For a moment, neither of them spoke.

Finally, her mom continued. "When I was younger, pressure mostly stopped when you came home."

Maya stared at the screen again. Now it followed people everywhere. Messages. Feeds. Notifications. Recommendations. Even silence online sometimes felt loud.

"I don't think parents always understand it," Maya admitted quietly.

Her mom smiled slightly. "**That's probably true**."

Then she added: "**But sometimes kids don't realize adults are trying to understand it while it's already changing again.**"

That stayed with Maya. Not because it solved anything. Because it felt honest.

Later that same week, Maya's phone buzzed with a message from her grandma.

"Sweetheart, can you help? I got an email saying my bank account is locked. It looks official. They want me to click the link and enter my information."

Maya's stomach dropped.

She showed the message to her mom.

Her mom sighed. "This is exactly what we worry about with your grandparents. Seniors get targeted a lot. Scammers use technology to make messages look convincing. They make them sound urgent and official."

Maya remembered what Leo had said about phishing. It wasn't just random spam anymore. The messages felt real because they were designed to feel trustworthy.

She called her grandma right away. "Hi Nani, don't click anything. I'm coming over."

When Maya arrived, her grandma looked relieved but embarrassed. "I almost did it," she admitted. "It even had my old address."

They sat together at the kitchen table and looked through the email carefully. The logo looked real. The wording sounded professional. Everything about it seemed designed to stop someone from thinking and start them reacting.

They checked the sender's address. It was slightly different from the bank's real one. Then they looked up the bank website themselves instead of using the link.

No account problem. No warning.

A few minutes later, her grandma's phone rang. The caller ID displayed the bank's name. A friendly voice explained that there was a problem with the account and asked her to confirm some information.

Maya listened carefully. The caller sounded calm. Professional. Confident. But something felt wrong.

The caller was asking for information that the bank should already have.

Maya shook her head. Her grandma hung up.

They called the bank's real number directly. Everything was fine.

Her grandma let out a long breath. "I feel so silly."

"You're not silly," Maya said quietly. "They make it hard to tell. That's the point."

Her grandma looked at the email again. "It sounded so convincing."

Maya nodded. "So did the AI answers..."

Her grandma smiled. "That's not exactly comforting."

Maya laughed softly. "No. But it does remind me of something."

"What?"

"**Just because something sounds confident doesn't mean it's right**."

That evening Maya looked back at something she had written earlier.

The more confident a system sounds, the less people question it.

Maybe that wasn't only true for AI. Maybe it was true for people too.

Before leaving, Maya helped her grandma write a simple reminder:

Pause. Verify. Ask questions. Then trust.

Maya watched her tape it beside the computer. It felt small. But it felt like something real they could do.

She left feeling a little lighter. Not everything was solved. But some things were clearer now.

That night Maya's mom set her phone on the kitchen counter. "New rule. Phone stays downstairs after nine."

Maya froze. "You're serious?"

"I'm serious. You've been up late scrolling. I worry."

"You don't get it," Maya said, voice rising. "Everything happens at night online. School stuff. Friend stuff. Everything."

Her mom crossed her arms. "I'm trying to understand, Maya. But I'm still your parent."

The words hung between them. Maya felt her face get hot.

She grabbed her phone and headed upstairs without saying another word. In her room she flopped onto the bed, staring at the ceiling.

Ten minutes later there was a soft knock. Her mom stood in the doorway.

"I'm not trying to punish you," her mom said quietly. "I just don't want you to burn out."

Maya didn't look at her at first. Then she sat up. "I know. I'm sorry I snapped."

Her mom nodded. "We'll figure it out together."

The door closed gently. Maya lay back down, phone still in her hand. For the first time in a while, she didn't open it.

CHAPTER 19

BUILDING THE MODEL (REAL SCENE)

The next afternoon Leo opened his laptop with fresh energy. "Let's build it properly this time," he said. "Not just play with it."

Maya leaned forward. "What's different now?"

"We decide what matters," Leo replied.

He created fields on the screen –

grades, activities, attendance, hours spent studying, even a small box for personal notes.

As they worked, Maya kept thinking about her own application and how neatly her life had been reduced to checkboxes.

"Run it," Jay said, excited.

The result appeared almost instantly.

Maya stared at the score for a moment, then lowered the attendance number slightly. The prediction dropped noticeably. She tried adding a strong activity. The result shifted again.

"So this isn't just about the data you put in," she said quietly.

Leo nodded. "It's about the decisions people made before the data even arrived."

Jay looked surprised. "Most people never see that part, do they?"

Maya thought about her rejected application - all those neat boxes and summarized lines. Two people with similar numbers could end up with very different outcomes depending on what the system valued most.

She adjusted the sliders herself. The rankings changed dramatically.

For the first time she understood something important.

The power wasn't only in the data. It was also in the invisible choices behind the data.

"Small changes create big differences," she said quietly.

Leo nodded. "Small choices inside a system can create big differences outside it."

Maya looked back at the rankings. A few adjustments had completely changed who appeared at the top.

The model wasn't simply describing reality. It was influencing how reality would be interpreted.

"Models don't just reflect reality," she said.

Leo smiled. "They shape it."

Instead of reaching for her notebook right away, Maya sat back and let the realization settle. She had spent weeks trying to understand why her application was rejected. Now she saw one possible reason more clearly than before.

Later she opened her notebook and wrote only one line:

The numbers are only part of the story.

Someone still decides what those numbers mean.

Small choices inside a system can create big differences outside it.

She closed the notebook. For the first time she felt she was beginning to see the systems differently - not as magic, but as choices made by people.

CHAPTER 20

WHAT DATA REMEMBERS

That evening, Maya opened her phone while finishing homework. Something felt different.

Earlier that week, she had spent hours researching scholarships, college programs, and volunteer opportunities. She had clicked, watched, and lingered - falling into one of those rabbit holes that quietly swallowed an entire evening.

Now her screen seemed filled with them.

A scholarship article sat near the top of her feed. Then a video about college admissions. Then an ad for student leadership programs.

She paused, thumb hovering.

A few days ago, none of this had been there. Now it felt like her phone had quietly made up its mind about who she was.

"Why am I seeing this?" she asked.

Jay leaned over. "You were just talking about scholarships."

Maya frowned. "Talking about them is one thing. How does my phone already know?"

Leo leaned forward. "What permissions did you give the app?"

They opened the settings.

Tracking: On.

Personalization: On.

Activity sharing: On.

Jay laughed. "So it knows everything."

Leo shook his head. **"It doesn't know everything. But it knows enough to recognize patterns."**

Maya leaned back. "That's still a lot."

She scrolled again. More suggestions appeared. Closer to what she had been thinking about earlier.

"It feels like it's following me," she said.

"It's following patterns," Leo corrected.

Maya didn't respond right away. Something still didn't sit right.

"How is it updating so fast?" she asked.

Jay shrugged. "It's just quick."

Leo shook his head. "It's not happening only on your device. It's running somewhere else."

Maya looked at him. "Somewhere else?"

Leo nodded. "On remote servers. What people call the **cloud**."

Maya sat back slowly. "So, every time I search… type… click…"

Leo nodded.

"And not just that."

Maya looked at him.

"What else?"

Leo took a moment.

“Think about everything around you,” he said. “Your phone tracks location. Apps track what you open. Watches track steps… heart rate.”

Jay frowned. “That’s still just devices.”

Jay pointed toward the window. “So where does all this data come from?”

Leo thought for a second. “Everywhere.”

“That’s a little creepy.”

Leo smiled. “Traffic lights.”

“Seriously?”

“Stores too.”

Maya looked up. “Stores collect data?”

“You’ve never searched for something once and then started seeing it everywhere?”

Maya stopped. Actually… she had. More than once.

Jay noticed immediately. “That face means yes.”

Data wasn’t just something people typed. It was something they left behind.

Maya blinked. “So data isn’t just what we type.”

“It’s what we do,” Leo said. “And sometimes… what we don’t realize we’re doing.”

Jay frowned at his phone. “So basically, my refrigerator probably knows more about me than I do.”

Leo stared at him. “That’s not exactly what I said.”

“Feels close enough.”

Later that same evening, Maya's phone buzzed with an alert.

Possible Fraud Detected on your card.

Jay looked at the alert. "Honestly, if I ever get one of those, it'll probably be because I bought food."

"Again?" Maya asked.

Jay looked offended. "I need support during stressful times."

"You were stressed during lunch."

"Exactly."

She stared at the notification. For a second, Maya thought someone had stolen the card. Her mom had used the card for groceries earlier. The system had flagged it instantly.

Maya showed the alert to Leo the next day. "See?" , "It caught something suspicious before we even knew."

Leo nodded. "That's **fraud detection**. It compares what's happening to what normally happens. Location. Timing. Spending patterns. If something suddenly looks different, it flags it."

Maya thought about it quietly. "So even when it helps us," she said, "it's still watching every pattern."

Leo didn't disagree. "Insurance works the same way sometimes. Companies look at driving patterns or other data to set your premium. One small change in behavior can quietly raise or lower what you pay."

Maya sat quietly for a moment. **Even ordinary habits could become signals now.**

Later that evening, she mentioned it to her dad while he was paying bills at the kitchen table. "So a score today can affect things years later?" she asked.

Her dad nodded. “That happens with credit scores too. Miss a payment, and the impact doesn’t always disappear right away.”

He showed her a simple example on his phone. “Interest works the same way. It compounds. **Small choices don’t always stay small. They build over time**.”

Maya watched the numbers grow. Patterns worked that way too. One decision became another signal. One signal became part of a larger story. Sometimes the consequences weren’t obvious until much later.

Her dad was still sitting at the table, comparing options for something that seemed important.

Maya noticed her dad looking at two long documents on his laptop.

"What are those?"

"Mortgage options," he replied.

She looked at the screen.

"They all look the same."

"That's what I thought."

He smiled and opened an AI tool.

Maya watched as he typed:

Compare these mortgage options and summarize the key differences.

A few seconds later, a summary appeared comparing interest rates, closing costs, monthly payments, and other trade-offs.

Maya looked surprised.

"That would've taken forever."

"It would've taken me a lot longer," her dad admitted.

Maya would've called it a shortcut.

But this felt different.

Her dad wasn't letting the AI decide.

He was using it to understand faster.

"Do you trust it?" Maya asked.

"Not completely."

He checked the numbers himself.

"Then why use it?"

Her dad smiled.

"Because understanding something quickly is useful."

He paused.

"But understanding it correctly still matters."

Maya thought about that.

The AI wasn't replacing the decision.

It was helping someone make one.

That felt different from everything she had seen before.

She opened her notebook.

The more a system knows, the more it can help. And the more it can influence.

What we do online doesn't disappear. It becomes part of the pattern.

Some choices feel small when we make them. But systems remember them longer than we do. Sometimes, they remember them before we realize they matter.

She wasn't just using the system anymore. She was starting to think about what it remembered.

CHAPTER 21

THE FILE YOU NEVER SEE

The next day after school, Jay invited them over. His father was working at the kitchen table, reviewing something on his laptop.

Jay dropped his bag and pointed at the screen. "Dad, show them."

His father looked up and smiled. "You want to see the invisible file?"

Jay nodded. "Yeah. The one that decides everything."

Maya and Leo gathered around.

"So there's a file about me?" Jay asked.

His father nodded. "Yes. A **credit file**. Most people have one."

"What kind of file?" Maya asked quietly.

"A record of how you've handled money—or how someone in your family has," he explained. "Lenders look at it before they approve loans, cards, even apartments sometimes."

Jay frowned. "So decisions depend on something I don't even see?"

His father didn't sugarcoat it. "That's right. It can feel unfair. Just invisible."

Maya looked at the screen.

The entry seemed insignificant.

But so had many of the signals they had talked about before.

Attendance.

Clicks.

Recommendations.

Patterns.

Small things mattered because systems remembered them.

Sometimes longer than people did.

"Then it needs to be corrected," his father said. "But first, someone has to notice."

Jay refreshed the page. "Wait… this doesn't look right."

They all leaned closer. An old entry was outdated.

"That's small," Jay said.

His father shook his head. "It can still matter. Small mistakes in these files can raise your rates or get you denied later."

Maya looked at the screen. She thought about her own future—student loans, renting a place, maybe buying a car one day.

"So something small…" she said softly.

"…can affect a much bigger decision," Leo finished.

Maya flipped to a new page in her notebook and wrote:

Decisions depend on data. Data depends on accuracy. And accuracy depends on attention.

She looked up, the weight of it settled in. **The invisible file suddenly felt very real.**

She thought about her own applications. Her own future. How much of it might already be sitting in files she would never see.

She closed the notebook. But the thought stayed open.

CHAPTER 22

THE FEED THAT DOESN'T STOP

Maya didn't notice when it started. It was just a few minutes of scrolling between tasks, watching one video and then another. By the time she looked up, an hour had passed.

Jay laughed. "That happens to me all the time."

Maya frowned. "It doesn't feel like an hour."

Leo glanced at her screen. "It's not designed to feel like time."

Maya looked at him. "What do you mean?"

Leo leaned back. "Most of these apps are built to keep you there."

Jay nodded. "Yeah, but it's just content."

Leo shook his head. "It's not random content."

Maya scrolled past something quickly, then went back without knowing why. Leo pointed out that **what appeared next was based on what she had just watched**. Jay scrolled again and realized the system was learning what he liked and giving him more of it.

"It's strange," Maya said. "We're told to trust systems because they're accurate, and at the same time we're told not to trust what we see."

She looked back at her screen and made a small decision. She closed the app. Not because she was angry, but because she wanted to see if she could choose when to stop.

That Friday night at the school fundraiser, music played softly. Students moved between tables and photo booths. Almost everyone had a phone out. Some barely looked at each other before taking pictures. One group stopped just to retake the same photo because the lighting looked bad. Another student checked the post right after uploading it.

Maya watched quietly. The real moment disappeared in seconds. Not fake. Just… filtered.

Jay glanced around. **"It's weird. Sometimes it feels like people enjoy the post more than the actual event."**

Later that night at home, Maya's younger cousin sat curled up on the couch watching short videos. Swipe. Laugh. Swipe again. The videos changed instantly, faster than conversations could settle.

"What are you watching?" Maya asked.

"I don't know," the kid answered without looking up. "It just keeps giving me stuff."

Maya stood there for a moment. Another video started before the last thought even finished.

She didn't write anything down immediately. Instead, she simply turned away from the screen and sat with the feeling.

Some systems weren't trying to inform her. They were trying to hold her attention.

Attention had become something valuable.

Valuable enough that entire systems competed for it.

CHAPTER 23

WHAT IT SHOWS NEXT

Maya opened the app again. Just to check. The feed looked familiar. Almost too familiar.

“It’s still the same,” she said.

Jay leaned over. “Yeah. That’s what you like.”

Maya frowned. “That’s what it thinks I like.”

Leo looked up. “That’s closer.”

Maya turned to him. “So how does it decide?”

Leo didn’t rush. “It doesn’t decide all at once.”

Maya waited.

“It builds over time,” he continued. “Small signals. Repeated patterns.”

Maya looked at her screen again. “What kind of signals?”

Leo pointed at Maya’s screen. “Try something.”

“What?”

“Watch that one a little longer.”

Maya let the video play. Then she scrolled. The next video looked strangely familiar.

Jay laughed. “Okay. That’s creepy.”

“I didn’t even like it,” Maya said.

Leo shrugged. “You stopped.”

The system wasn’t waiting for opinions. It was watching behavior.

Jay nodded. “So even if I don’t like something…”

“It still learns from it,” Leo said.

Maya blinked. “So just watching is enough?”

“Sometimes,” Leo said.

Maya leaned back. “That’s… a lot.”

Leo nodded. “**It doesn’t need everything. Just enough patterns to predict what you’ll do next.**”

Maya thought about that. “Predict,” she repeated.

Leo nodded. “It’s always guessing what you’ll stay for.”

Maya looked at her screen again. “So it’s not showing what’s best.”

“It’s showing what works,” Leo said.

Maya scrolled once more. The next video appeared instantly. No delay. No choice. Just… next.

“It already decided,” she said quietly.

Leo nodded. “Before you asked.”

A few days earlier, Maya had found herself stuck in a group project with Aarohi and two other students.

The assignment wasn’t difficult. The people were.

“We should just split the work and combine it later,” one student said.

“That’s how we end up with four completely different presentations,” Aarohi replied.

Maya quietly agreed with Aarohi but had her own idea. “What if we use real photos from the events instead of stock images? It would feel more honest.”

Aarohi barely looked up. “We don’t have time to change everything now.”

Maya stayed quiet. The frustration built in her chest. She stared at her notebook instead of arguing.

Later in the hallway Aarohi caught up with her. “Hey… sorry if I was bossy back there. I just get carried away sometimes.”

Maya shrugged. “It’s okay.”

“No, it’s not,” Aarohi said. “Your idea was good. We can still add some photos.”

Maya felt the knot in her stomach loosen. “Yeah?”

“Yeah.”

They walked the rest of the way in comfortable silence.

Aarohi was usually the first person to laugh during group projects. Not the loudest laugh, just the kind that made other people relax without realizing it.

Later that evening, Maya saw another post from Aarohi.

Perfect lighting. Perfect timing. Perfect caption. Everything looked effortless.

Maya stared at it a little longer. Then something about it started to feel familiar. Not fake. Just… filtered.

The next day, Maya saw Aarohi sitting alone near the hallway windows. No phone raised. No camera angle. No perfect smile. Just quiet.

Maya noticed Aarohi still carried an old water bottle covered in fading stickers from middle school. It looked strangely out of place next to the polished version of herself online.

For a second, Maya almost kept walking.

Then Aarohi looked up. “Hey,” she said.

“You okay?” Maya asked.

Aarohi gave a small shrug. “Just… tired, I guess.”

That surprised Maya. Online, Aarohi always looked ahead of everyone. Confident. Perfect. Finished. This version felt different. **Real.**

Aarohi glanced at her phone. “Sometimes it feels like if I stop posting,” she said quietly, “people think I disappeared.”

Maya didn’t answer right away. “That sounds exhausting.”

Aarohi laughed softly. But it didn’t sound like a joke. “Yeah,” she admitted. “But if you disappear too long… the algorithm kind of forgets you.”

The answer stayed with Maya. Not because it sounded dramatic. Because it sounded normal.

Aarohi locked her phone. “For what it’s worth,” she said quietly, **“most people online are struggling more than they look. They just don’t post that part**.”

Then she stood up. “See you later.”

Maya watched her leave. For the first time, Aarohi didn’t feel ahead. Just… **human**.

That evening Maya opened the app again. The feed kept moving. Another video. Another recommendation. Another perfect moment.

She thought about Aarohi. She thought about the game. She thought about how easily systems learned what people stayed for.

And how quickly people started adapting to what systems rewarded.

The screen wasn't forcing anyone to do anything. It didn't have to. It simply made some choices easier than others. Some voices were louder than others. Some versions of ourselves are easier to show than others.

Maya opened her notebook.

The feed doesn't just learn from people. People learn from the feed.

She stared at the sentence for a moment. Then added one more.

What gets attention grows.

She closed the notebook.

For the first time, she understood something important.

The system wasn't only shaping what people watched. It was quietly shaping what people noticed. What they shared. What they valued. And, little by little, what they became.

CHAPTER 24

THE PRICE OF ATTENTION

Maya opened the app again. It didn't wait. It already knew what to show next.

She looked at the screen for a moment. "So it decides before I do," she said.

Leo nodded. "And then it nudges what you do next."

She didn't plan to buy anything. She was just scrolling.

A video appeared. Simple. Clean. Someone talking about a product. "It changed everything," the voice said.

Maya kept watching.

The next video showed the same thing. Different person. Same product.

Jay leaned over. "Didn't you just see that?"

Maya nodded. "Yeah."

But she didn't scroll away. Something about it remained.

Later that evening, she opened a shopping app. The product was there. At the top. "Recommended for you."

She took a moment. "I didn't search for this," she said.

Leo looked at her screen. "You didn't need to."

Maya frowned. "It connected the signals," Leo said.

Jay leaned back. “So it followed her from one place to another.”

Leo nodded.

Maya checked the screen again. “It feels like it already decided what I want.”

Leo didn’t disagree. “Sometimes systems don’t wait for decisions. They guide them.”

“It’s not just showing options,” Leo said. “**It’s shaping decisions**.”

Maya hesitated. “What if I didn’t actually need this?”

Jay smiled. “But now you want it.”

That part felt uncomfortable. Because it was true.

Maya didn’t answer. She closed the app. Then opened it again. The product was still there.

Maya looked at the screen one last time. It didn’t feel random anymore.

Jay looked at Maya.

"You know you've started doing what Leo used to do."

"What?"

"Finding weird stuff for us to investigate."

Maya laughed.

The notebook pages are turned later that evening, she wrote

What we see can shape what we choose. And sometimes, it shapes it before we notice.

She closed the notebook slowly. She wasn’t just thinking about what she saw. She was thinking about what it made her do.

CHAPTER 25

THE WORLD THAT LOOKS PERFECT

The video started like the others.

Bright colors. Clean spaces. Perfect timing.

Someone laughing. Traveling. Making it look effortless.

Jay watched it twice. “That looks amazing.”

Maya nodded. “It does.”

Leo didn’t react the same way. “What do you notice?” he asked.

Maya looked again. Everything felt smooth. **Too smooth.**

“There’s nothing wrong in it,” she said.

Leo nodded. “That’s part of it.”

Jay frowned. “What do you mean?”

Leo leaned back. “You’re seeing the best version.”

Maya scrolled again. Another post appeared.

Jay paused. “Do you follow Aarohi?”

Maya nodded.

“She’s doing everything,” he said. “Internships, scholarships, all of it.”

Everything looked perfect. Maya zoomed in. There was something slightly off. Too perfect. It felt complete. Like nothing was missing.

That was what made it hard to question.

She took a moment. “It’s not just the videos,” she said quietly. “It’s people too.”

Leo didn’t disagree.

Maya kept scrolling.

Another perfect post.

Another success story.

Another carefully framed moment.

She stopped for a second.

"What if this isn't the whole picture?" she asked.

Jay looked over.

"What do you mean?"

Maya turned the screen toward him.

"Everything looks perfect."

Leo nodded.

"That's usually the point."

Maya looked at the image again.

For the first time, she wasn't thinking about what she was seeing.

She was thinking about what might be missing.

The hard days.

The doubts.

The failures.

The parts nobody posted.

Maya kept scrolling. More success. More speed. More results.

"Make money fast." "Anyone can do this." "Start today."

Jay leaned forward. "That sounds real."

Maya hesitated. "**Or it's designed to feel real**."

Leo nodded. "Some content is built to create urgency. To make you feel like you're behind."

Maya glanced at the screen. It didn't feel inspiring anymore. It felt pressuring.

She turned the screen off. The room felt quieter.

She sat quietly then wrote:

What looks perfect is often incomplete. Some systems don't show reality. They show what keeps attention.

Across the room, someone laughed. A phone buzzed. Another post. Another story.

Maya didn't reach for her phone. For the first time, she wasn't just thinking about what she was seeing. She was thinking about what she wasn't.

CHAPTER 26

THE ANSWER BEFORE THE QUESTION

Maya noticed it during something simple: homework.

Jay opened his laptop, typed a question, and an answer appeared instantly. He started copying it straight into his assignment without even pausing to read it fully.

"Wait," Maya said.

Jay looked up. "What?"

"You didn't even think about it," she said.

Jay shrugged. "It's already there. And it's correct."

Maya frowned. "That's not the point."

Leo glanced over. "It is correct."

Maya nodded slowly. "But you didn't figure it out yourself."

That sentence stayed with her.

The next day at lunch she watched the cafeteria differently. Groups of students sat together, but few were really talking. Most were staring at their screens, scrolling, watching short videos, tapping replies. Between classes the hallways looked the same: eyes down, thumbs moving, people walking but not really present.

She remembered evenings at home when the same thing happened. Phones on the dinner table. Screens lighting up between bites. Conversations starting, then fading as soon as a notification arrived. It all felt completely normal now.

Maya turned to Leo and Jay.

“Do you notice that?”

Jay glanced around.

“They’re just on their phones.”

Maya shook her head.

“No. It’s more than that.”

Nobody seemed stuck anymore. If someone didn’t know something, they looked it up immediately. Nobody waited. Nobody wondered. Nobody sat with a question for very long.

Leo nodded.

“I think you're noticing something important. If answers are always available, you stop asking people as much. You stop struggling through problems together. You stop waiting to hear what someone else thinks.”

That changed how it felt.

A few nights later, Maya stood awkwardly near the edge of the gym floor at the school dance. Lights flashed. Music thumped. Most kids had their phones out, posing for pictures or filming short videos.

Jay nudged her.

“You gonna post anything?”

Maya shrugged.

“I don’t know. Everything looks better online than it feels right now.”

Leo stood beside them, watching the crowd.

“That’s kind of the point, isn’t it?”

Maya looked at the sea of glowing screens. For a second she felt the pull, the desire to capture a perfect moment so it could live online. Then she put her phone back in her pocket.

She danced anyway.

Not for the feed.

Just for the music.

Later she wrote in her notebook:

Sometimes the best moments happen when no one is watching.

Later that week she overheard two seniors in the hallway talking about using AI for internship reports. One said it got everything done in minutes. The other laughed and admitted he mostly just checked the output.

That afternoon Maya saw Aarohi in the hallway—not on a screen, but real. Aarohi was walking quickly, checking her phone between steps. She looked tired.

“You okay?” Maya asked.

Aarohi gave a small, real smile.

“It’s been a lot lately.”

They stood there for a moment. Aarohi admitted she sometimes spent more time thinking about what to post than actually doing the thing. And **if she stopped posting for too long, it felt like the algorithm forgot her.**

Maya didn’t know what to say.

For the first time, Aarohi didn’t feel ahead.

She felt human.

Maya sat down on a bench nearby.

A notification buzzed somewhere.

Then another.

Nobody looked up.

The answers were everywhere, but she found herself wondering whether people were asking fewer questions.

Instead of writing everything down, Maya sat quietly on the bench.

She wrote:

I don’t have to let the tools think for me.

I can still choose my own thoughts.

CHAPTER 27

THE QUESTION EVERYONE ASKS

The technology lab was nearly empty. A few monitors still glowed in the corner. Someone had left a half-finished robotics project on a table near the window.

Jay dropped into a rolling chair and spun once.

Leo immediately looked up. "You're going to break something."

Jay grinned. "Relax. If anything breaks, we'll blame Maya."

Maya looked up from her screen. "Why me?"

"Because nobody ever suspects the quiet one."

Maya laughed.

Suggestions.

Auto-complete.

Recommendations.

Things appeared before she even asked for them.

Jay typed a sentence. The rest filled itself in. He didn't even pause before accepting it.

Maya watched. "It already knows what comes next."

Leo glanced over. "It predicts."

"That feels like deciding."

"It's suggesting," Leo replied. "But suggestions influence decisions."

Jay leaned back. "So it's helping."

"Sometimes."

Maya looked at the screen again. "It's also shaping."

Leo nodded. "That's the part people don't always notice."

Maya's phone buzzed during the lab. She quickly typed a reply to Jay: "This AI stuff is kind of stressing me out lol."

She hit send without thinking.

Jay's reply came fast: "Wow. Didn't know I was stressing you out."

Maya stared at the message. "Wait, that's not what I meant."

She typed faster. "I meant the topic. Not you."

Jay didn't reply right away. When he finally did, it was just: "K."

The rest of the lab felt awkward. Maya kept glancing at him.

At the end of class she caught him by the door. "Hey. I didn't mean it like that. The AI conversation is just a lot sometimes."

Jay looked relieved. "Oh. I thought I was being annoying or something."

"You're not," Maya said. "You're the opposite."

Jay grinned. "Good. Because I bring the energy, remember?"

They both laughed. The awkwardness disappeared as quickly as it had arrived.

Jay clicked a link. Nothing happened. He clicked again. Still nothing.

"Seriously?"

Leo shook his head. "Not exactly. It runs somewhere else. Not on your phone."

Jay frowned. "Where then?"

Leo pointed toward the window. "Probably hundreds of miles away."

Jay looked outside. "There?"

Leo laughed. "Not that building."

Jay smiled. "Good. That would've been disappointing."

"Big data centers," Leo said. "Thousands of servers. Processing. Storing. Predicting."

Maya considered that. "So everything here connects to something much bigger somewhere else."

Leo nodded. "Pretty much."

The room went quiet for a moment.

Jay stared at his phone. "That's actually less magical and more weird."

"Pretty much," Leo said again.

Later that week, Leo showed them something new on his laptop.

“Watch this,” he said. “This AI doesn’t just answer questions. It can actually do things—open tabs, write emails, even book appointments if you tell it to.”

Maya frowned. "So, what's the difference?"

Leo opened his laptop.

"Imagine I ask a regular AI to plan a vacation."

"It gives you ideas," Jay said.

"Right," Leo replied. "Now imagine I tell an agentic AI: Find the cheapest flight. Compare hotels. Pick one near the beach. Book it. Then email me the itinerary."

Maya blinked.

"It can do all of that?"

"That's the goal," Leo said. "Instead of just answering questions, it can break a goal into steps and complete the task."

Jay laughed.

"So basically a digital assistant that never sleeps."

"Something like that," Leo said.

Maya looked thoughtful.

"That sounds useful."

"It can be," Leo replied.

Maya leaned forward.

"What happens if it misunderstands what you want?"

Leo smiled.

"That's one of the biggest challenges."

Jay looked confused.

"Why?"

"Because a system can follow instructions exactly and still do the wrong thing."

Maya raised an eyebrow.

"How?"

Leo thought for a moment.

"Imagine telling it to find the cheapest hotel."

"Okay."

"It finds one that's miles away from everything, has terrible reviews, and no transportation."

Jay laughed.

"It technically succeeded."

"Exactly," Leo said.

"**It completed the goal. Just not the goal you actually had in mind**."

Maya nodded slowly.

"So getting the answer right isn't enough."

"It's about understanding the intent behind the answer."

Jay whistled softly.

"So we're basically teaching computers how to complete tasks now?"

“Kind of,” Leo said. “But it’s getting smarter. Some people are building systems that can work on goals by themselves. Plan steps. Use tools. Keep going until the task is done.”

Maya leaned in. “So, it doesn’t wait for us to click every button?”

“Exactly. It can act.”

For a moment the lab felt quieter.

“That’s… powerful,” Maya said slowly.

Leo nodded. “Yeah. But it also means we have to be even more careful about what we ask it to do. And who we let it act for.”

Maya thought about that on the walk home. Systems that didn’t just suggest anymore. Systems that could do things on their own.

Maya looked around the room. **A few months ago, she thought AI was something that happened on screens. Now it felt like something quietly moving through decisions, recommendations, conversations, and choices every day.**

She opened her notebook and wrote one line:

Some tools don't just help us make decisions. They can start making them too.

Leo continued. “Some people call the next step AGI. **Artificial General Intelligence**.”

“Meaning?” Jay asked.

“A system that could learn, adapt, and reason across lots of different kinds of problems.”

Maya leaned back slightly. “That sounds… far away.”

“Maybe,” Leo said. “But people used to say the same thing about systems making decisions for us.”

"Do you think it'll eventually be able to do almost anything?"

Leo shrugged.

"Maybe someday. Nobody really knows where the limits are."

Leo shrugged. “Honestly? Nobody really knows.” He glanced toward the robotics project. “But we’re definitely not there yet.”

A few days later, Maya was helping at an after-school technology event for younger students.

Most of the questions were simple. How do robots move? How does AI make pictures? Can computers think?

Maya answered what she could and pointed the harder questions toward the teachers.

As the event was ending, one younger student lingered near her table. He looked nervous.

“Can I ask something?” he said.

“Sure,” Maya replied.

The student hesitated. “How do you know if an AI answer is right?”

Maya smiled. A year ago, she would have answered very differently. Now she took a moment.

“You don’t always know.”

The student looked disappointed. “So then what’s the point?”

“The point isn’t to trust everything it says,” Maya replied. “The point is to use it to help you think.”

The student frowned slightly. “That’s confusing.”

Maya laughed softly. “Yeah. It confused me too.”

She thought for a moment. “If a calculator gives you an answer, you still need to know what problem you’re solving. AI is kind of like that. It can help. It can explain things. It can give ideas. But it shouldn’t replace your own thinking.”

The student nodded slowly. "So I should still ask questions?

"Especially then," Maya said.

The student smiled. "Okay. I can do that."

As he walked away, Maya realized how different that answer would have sounded a few months ago.

At the beginning of the year, she had wanted systems to give her answers. Now she understood that the better skill was knowing which questions to ask.

That night, Maya sat at her desk with her laptop open. She had a report due the next day. For a moment she stared at the blinking cursor.

She could ask the AI to write most of it. It would sound good. It would be fast.

Instead, she started typing herself. Slowly at first. Then a little faster. She paused, deleted a sentence, and rewrote it. It felt harder. But it also felt like hers.

She smiled quietly to herself.

She wasn't just learning how systems worked anymore. She was learning how to work with them - without letting them do everything for her.

She wrote in her notebook:

I can choose when to use the tools. And I can choose when to do it myself. That choice matters.

The lab felt quieter now. But inside Maya's mind, the questions kept growing.

CHAPTER 28

THE HARD PART TO SEE

Maya didn't expect it to become real. At first, everything online felt distant. Stories. Videos. Messages. Things that happened to other people.

Until one message didn't.

It came late. Simple. Familiar.

"This deal ends soon—don't miss it."

The name looked right. The profile looked normal.

Jay leaned over. "You know them?"

Maya hesitated. "I think so."

Leo didn't look convinced. "Look carefully," he said.

Maya opened the message again. Something felt slightly off. The wording. The timing. The link.

"It's close," she said.

She took a moment. It reminded her of something. Aarohi. The posts had looked complete. Convincing. Easy to believe. But they hadn't shown everything.

She looked back at the message. This felt the same. Not everything that looks complete is safe to trust.

Leo nodded. "That's usually enough."

Jay frowned. “What happens if you click it?”

Leo didn’t exaggerate. “Sometimes it asks for information. Passwords. Details.”

Maya leaned back. “And if someone enters it?”

Leo answered quietly. “Then they’re not the only one using it anymore.”

That wasn’t confusing anymore. It was real.

A few days later, something else happened. Jay looked frustrated.

“What?” Maya asked.

He showed his screen. “I bought something. It looked real.”

Maya leaned closer. The site looked clean. Professional. Convincing.

“What happened?”

“It never arrived,” Jay said.

Leo nodded slowly. “That happens more than people think.”

Maya sat back. “It wasn’t obvious.”

Jay shook his head. “That’s the problem.”

Maya thought about the message again. The purchase. The videos. None of them looked dangerous. That was what made them dangerous.

She didn’t want the thought to be forgotten ...

Not everything online is meant to help.

She took a moment. Then added:

Some things are designed to look real, create urgency, or gain trust quickly.

The danger isn't always obvious. That's what makes it hard.

She read the lines again, this time, it didn't feel distant anymore.

CHAPTER 29

THE GAME THAT DOESN'T END

Nothing felt urgent anymore. But something still felt different.

The next day felt normal. That was the part Maya noticed. Nothing urgent. Nothing obvious. Just… routine.

Jay sat down and opened a game. "Just one round," he said.

Maya didn't react. She had heard that before.

The game loaded quickly. Bright colors. Fast movement. Immediate feedback.

Win.

Jay smiled. "That was easy."

He tapped again. It didn't feel like a decision. That was the part she noticed. Another round.

Maya glanced at the screen. "You just started," she said.

Jay shrugged. "It's quick."

Leo looked over. "That's part of it."

Maya tilted her head. "What do you mean?"

Leo leaned back. "Some systems are designed to keep you coming back."

Jay didn't look away. "It's just a game."

Leo nodded. “Yes. But it’s also a system.”

Maya watched the screen. Win. Lose. Almost win.

Jay leaned forward. “That was close.”

He tapped again.

Maya noticed something. It wasn’t the game. It was the timing. The near wins. The quick retries. The instant restart. It didn’t give space to stop.

“Do you feel like stopping?” she asked.

Jay paused for a second. Then shook his head. “Not yet.”

Leo nodded. “That’s the loop.”

Maya looked at him. “What loop?”

Leo spoke calmly. “Action. Reward. Repeat.”

Maya watched another round. It made sense. Every action led to something. Points. Progress. Feedback.

“Even losing doesn’t stop it,” she said.

Leo nodded. “It keeps you engaged.”

Jay leaned back briefly. “Okay, one more,” he said.

Maya didn’t say anything. She just watched. The game wasn’t forcing him. It didn’t need to. No pressure. No urgency. Just… continuation.

It was guiding him. Round after round.

Maya thought about the message from before. That had felt urgent. This didn’t. But it still kept going.

“That’s different,” she said.

Leo looked at her. “How?”

Maya thought for a moment. “That pushed me to act fast. This makes you stay.”

Leo nodded. “Different systems. Same goal.”

Jay paused the game. For a moment. Then resumed.

Maya noticed that too. It wasn’t a big decision. But it kept happening.

She took a deep breath:

Some systems don’t push you. They pull you.

They don’t need urgency. They use repetition.

And **the hardest patterns to see are the ones we don’t question.**

She looked at Jay. He was still playing. Not stressed. Not rushed. Just… continuing.

She noticed something new. She wasn’t just thinking about what systems showed her. She was thinking about how they shaped what people kept doing.

CHAPTER 30

ASKING WHY

Maya stared at the screen longer than usual. The result still hadn't changed. She closed the laptop halfway, then opened it again.

"This doesn't make sense," she said quietly.

Jay leaned over. "It doesn't have to make sense to you. It just has to match what the system learned."

Maya shook her head. "That's exactly the problem."

Leo looked at her carefully. "What are you thinking?"

Maya took a slow breath. "I want to know why."

Jay gave a soft laugh. "You really think they'll tell you?"

"I don't know," Maya replied. "But I haven't even asked yet."

For a moment no one said anything. Then Leo nodded. "That alone changes things."

Only a few days remained before the appeal deadline. For weeks, Maya had been collecting questions about patterns, bias, predictions, and fairness. But when she opened the appeal form, she stopped. The real problem wasn't that she disagreed with the decision. It was that she still didn't understand it.

So instead of writing the appeal immediately, she opened a new message. If she was going to challenge the decision, she wanted to understand it first.

Later that evening, she drafted a careful note. Not emotional. Just clear. "Could you please explain which factors most influenced this decision?"

She read it several times and hit send.

The reply came the next day.

Polite. Structured. Careful. And very vague. "Your application was evaluated based on multiple factors…"

Maya read the response twice, then a third time. A few months earlier she would have stared only at the outcome. This time she found herself looking for something else—what data had been used, what had been ignored, who had reviewed it.

The answer still wasn't there. But the question had changed. And somehow that felt important.

Jay leaned in. "That's it?"

"That's it," Maya said.

Leo nodded slowly. "It's not designed to explain. It's designed to close the conversation."

Maya sat back. The frustration returned, but this time it felt clearer. Sharper. Not everything unclear was random. Some things were hidden on purpose.

That night she sat by herself in her room. No videos played in the background. No notifications interrupted her thoughts. For the first time in a while, the quiet felt intentional.

She thought about all the invisible systems they had explored—patterns, predictions, black boxes, deepfakes, attention loops. It wasn't that the systems made no sense. It was that most people never got to see how they really worked until the systems affected them directly.

Jay looked frustrated. "So that's it?"

Maya looked down at the notebook she had carried for months. Pages filled with questions she had never noticed before. She closed it quietly.

“I wanted an explanation,” she said. “But maybe the bigger problem is how many people never get one.”

For the first time, Maya didn’t just feel inside the system. She felt aware of it. And somehow that changed everything.

That night she opened a blank document. For a moment she just stared at the screen as the cursor blinked. Questions she had carried for months sat quietly beside her. The result had not changed. The explanation had never come. But the questions still mattered.

Maybe she couldn't change the decision. But she could refuse to stop asking questions about it.

She placed her notebook beside the keyboard and started writing.

CHAPTER 31

THE FACE THAT ISN'T REAL

Maya noticed something.

The video appeared while Maya was scrolling. At first she barely paid attention. Then she stopped. Something felt off.

The voice sounded right. The face looked real. But she couldn't explain why it felt strange.

She watched it again. Then a third time.

Finally she sent it to Leo and Jay. "Am I imagining this?"

A few seconds later, both of them replied.

Leo opened his laptop during lunch. "Watch this," he said quietly.

He played the same short video. A famous speaker stood on stage giving a powerful talk. The voice sounded exactly right. The movements looked natural.

Jay watched it twice. "I don't see anything wrong with it."

Leo paused the video. "Look closer."

Maya leaned in. Something felt slightly off. The blink rate. The way the mouth moved. It was close, but not quite human.

"That's not a real person," Leo said. "People call these **deepfakes**. They're created by AI."

Maya frowned. "The scary part isn't that it fooled me."

Jay looked up. "What is?"

"That I almost didn't stop to question it."

Maya felt a chill. "If this looks real… how do we know what's actually true anymore?"

Jay frowned. "So anyone can make this now?"

Leo nodded slowly. "The tools are getting easier to use. And they're getting better every month."

Maya thought about Aarohi's perfect posts. "What if someone used this to fake a message from a teacher? Or a parent? Or even a bank?"

Leo didn't sugarcoat it. "It's already happening. Some people create fake videos to spread false news. Others use them to trick friends or family."

Jay leaned back. "This means seeing is no longer believing."

"That's right," Leo said. "Seeing is no longer proof."

Maya stared at the frozen image on the screen. The face looked so real. Yet it was completely made up.

She imagined someone sending her a video of her mom asking for money or personal information. She would probably believe it.

The thought stayed with her.

Later that afternoon, she noted carefully:

Seeing is no longer proof. Verification is.

Reminder to self… In the future, trust will need more than eyes. It will need questions.

The library felt a little different now. Even the videos on her phone felt slightly less trustworthy.

She wasn't just thinking about deepfakes anymore. She was thinking about how easy it had become to create reality.

CHAPTER 32

THINKING LIKE A DEFENDER

The next day, Leo opened a different kind of screen.

"This is how systems are tested… most of the time," he said.

Jay leaned forward with interest. "Hacking?"

Leo shook his head. "**Ethical hacking**. There's a big difference."

Maya tilted her head. "What's the difference?"

"Permission," Leo said. "Purpose. And responsibility. These people have permission to find weaknesses so the systems can be fixed."

He showed them a practice system. "What would you try first if you wanted to break in?"

Jay thought for a moment. "Find a weak password?"

Leo smiled slightly. "That's one way. But good defenders think like attackers."

Maya watched quietly. "In order to protect something…"

"You have to understand how it can fail," Leo finished.

They spent the next twenty minutes trying simple attacks on the test system. Each time it failed, Leo explained why. Each time it almost worked, he showed how it could be improved.

Maya found herself thinking differently. "Not everything is built perfectly," she said.

"No," Leo agreed. "That's why testing matters. And why people who think like defenders are important."

Jay leaned back. "So even the best systems need humans watching them."

"Exactly," Leo said. "**Systems should guide decisions. Not replace the people who take responsibility**."

Maya took a moment then wrote:

Security isn't about avoiding problems. It's about preparing for them. Systems should help us. But they should never replace our judgment.

She closed the notebook and looked at Leo. "So, after everything we've learned…"

Leo waited.

"Should systems ever make big decisions completely on their own?" Maya asked.

Leo shook his head slowly. "They can help but not replace people. They can suggest. But the final choice should still belong to people."

Maya nodded. That felt right.

"Do you ever feel like it's getting harder to tell what's real?" Jay asked.

Maya nodded. "Not just real… but what to trust."

Leo leaned back. "That's why understanding matters now more than ever. Not just what systems do - but how they shape what we see."

“I learned most of this from books, online courses, and people sharing what they discovered.”

Maya looked at the screen, then away from it.

“For the first time,” she said quietly, “I feel like noticing it might actually change something.”

She didn’t feel powerless against the systems. She felt ready to question them.

She wrote: Systems should guide decisions. Not replace them. Understanding didn’t change everything. But it changed how she made decisions.

CHAPTER 33

THE CONVERSATION THAT MATTERS

Maya sat quietly in the library. Jay scrolled on his phone while Leo read something on his laptop. For a while nobody said anything. That wasn't unusual anymore. A few months ago the silence would have felt awkward. Now it felt comfortable, the kind that only happens around people who don't need to fill every moment with words.

Jay finally looked up. "You two are doing the weird silent-thinking thing again."

Maya smiled. "You noticed?"

"It's hard not to," Jay said.

Leo didn't even look up. "You're talking during it."

Jay sighed dramatically. "I bring energy to this friendship."

Maya picked up her phone. The feed loaded instantly - video after video with no pause. She watched one for a few seconds, then stopped. Not because it ended. Because she chose to stop. Notifications appeared. She ignored them. This time she decided when to respond.

Jay looked over. "You're done already?"

Maya nodded. "For now."

Leo glanced at her. "You see it now," he said softly.

Maya took a moment. "I think I do."

Jay leaned back. "Curious what changes?"

Maya thought about it. "The systems don't stop. They won't. But I can choose how I move through them."

Jay frowned. "What's the point then?"

Maya looked at her phone, then at her notebook. "It doesn't change what they do. It changes what I do."

Leo didn't say anything. He didn't need to.

Instead of writing immediately, Maya simply closed the app and sat with the quiet feeling. For a long time it had felt like something invisible was deciding for her. Now it felt different. Not because the systems had changed. But because she had.

A few days later, Jake stopped Maya in the hallway.

"Hey."

"Hey."

"I heard you're applying again."

Maya smiled. "Yeah."

Jake hesitated, then gave a small laugh. "For what it's worth… I almost didn't accept the scholarship."

Maya looked surprised.

Jake shrugged. "Turns out getting in doesn't magically solve everything."

For a moment neither of them said anything. Then Maya smiled.

"I guess we both learned that."

Jake nodded. Different paths. Different challenges. Neither of them fully knew where they were going. But neither was standing still.

That conversation stayed with Maya as she sat at the kitchen table later that evening. Another application form was open on her screen.

Her mom paused while walking by. “Another one?”

Maya nodded.

“A little nervous?”

“A little.”

Her mom smiled. “Good.”

“What?”

“It means you still care.”

Maya smiled and glanced toward the old volunteer T-shirt hanging over the back of a chair nearby. It had become just another piece of fabric she barely noticed. Yet it reminded her of all the weekends she had spent helping others, long before any scholarship application existed.

Months ago, success had meant getting selected.

Now she wasn't so sure.

She still wanted the scholarship.

But she wanted something else too.

To understand things.

To build things that help people.

To help people ask better questions.

For the first time, the future felt bigger than a single result.

She looked back at the screen. For the first time, nervousness didn’t feel quite the same as discouraged.

A few weeks later, another application deadline appeared. She stared at it for a moment. Months earlier she would have focused only on the result: accepted, rejected, selected, not selected.

Now she noticed different questions: What information would they see? What would they miss? Who would review it? How would decisions be explained?

She smiled slightly. Not because she had the answers. Because she knew what questions to ask.

Then she opened the application. For a moment she stared at the empty form. Months ago it would have felt like another judgment waiting to happen. Now it felt like an opportunity. Not guaranteed. Not certain. Just possible. And somehow, possibility felt better than certainty.

She started writing. This application felt different from the first one. Not because the system had changed. Because she had.

She thought about the first rejection again.

For months she had treated it like a verdict. A final answer. Proof that she wasn't good enough.

Now it felt different. The rejection still mattered. But it no longer explained who she was. It was one decision. Not her future.

She still hoped for a scholarship. She still cared about the outcome. But she no longer believed a single decision could define her future.

She stared at the application for a moment. The truth was part of her still cared about the original rejection. Maybe she always would. Not because she thought the decision was wrong. Because it had hurt.

For months she had treated that email like proof that she wasn't good enough. Proof that someone else had been chosen and she hadn't. Proof that the door had closed.

Now it felt different. The rejection still mattered. But it no longer explained who she was. It was one decision. One moment. One path she didn't take.

She thought about all the questions that decision had started. The books she had read. The conversations with Jay and Leo. The things she had learned about systems, people, and herself.

Without that rejection, none of it would have happened.

For the first time, she wondered if understanding had been worth more than the acceptance itself.

She thought briefly about Jake. For a long time she had treated his acceptance and her rejection as opposite outcomes. Now it felt more complicated than that. His path had challenges. Hers had questions. Neither of them had been defined by a single result.

Maya smiled slightly. Without the rejection, none of this would have happened. She would never have met these questions. Never have noticed the systems. Never have learned how often people confuse predictions with certainty.

The result she wanted had disappeared. But it had given her something unexpected: understanding.

Some opportunities require essays. Some require interviews. Some looked at different strengths. For the first time, **Maya understood that one result was not the same as her future. A system could influence a path. It could not define it.**

The future still felt uncertain. But it no longer felt decided.

Months earlier, Maya had wanted the system to tell her who she was. Now she understood something better. No score. No model.

No application. No algorithm. Could ever know her as completely as she knew herself. And that made all the difference.

There was still space. To learn. To create. To choose. To question.

And maybe that was the most important lesson of all. Not that systems would disappear. Not that every decision would become fair. Not that every answer would be visible. But that people could keep asking better questions.

Maya smiled and returned to the application. Then she started writing.

ONCE YOU NOTICE

What systems learn shapes, what they show.

What they show shapes, what we do.

What we do shapes, what comes next.

And most of the time, we don't even realize we're part of the cycle.

But once you notice, you don't move through it the same way again.

Maya never found perfect answers.

Most people probably don't.

But once she started noticing the systems around her, she couldn't completely stop noticing them anymore.

And maybe that wasn't a bad thing.

Because sometimes, the most important questions begin after the story ends.

Even as new systems appeared, ones that could plan and act on their own - Maya kept coming back to the same truth. The most powerful thing wasn't the technology. It was her ability to notice it, question it, and still choose for herself.

A FEW THINGS MAYA KEPT COMING BACK TO

By the end of the year, Maya's notebook wasn't organized. Some pages had questions. Some had observations. Some were ideas she still wasn't sure about.

But a few concepts kept showing up repeatedly:

Patterns, Not People

Systems don't know individuals. They recognize patterns from past data and apply them to new situations.

What's Missing Matters

Systems can only use the data they have. What isn't measured can't be considered. And what isn't considered can still matter.

Small Signals, Big Impact

Tiny inputs—like attendance, clicks, or timing - can influence outcomes more than expected.

Predictions vs Reality

Systems predict what is likely. But predictions are not guaranteed. They become decisions when people stop questioning them.

Feedback Loops

Systems don't just observe behavior. They can influence it—and make their own predictions come true.

Accuracy vs Fairness

A system can be accurate overall and still feel unfair in specific situations.

Trust and Verification

Seeing is no longer enough. Trust now requires questions, not just observation.

Maya never found perfect answers to all of these ideas. That wasn't really the point.

The point was noticing them.

Because once you notice something, you start asking questions.

And once you start asking questions, you begin to see things differently.

If YOU'RE CURIOUS: A CLOSER LOOK AT HOW IT WORKS

You don't need this section to enjoy the story.

It's here only if something in the book made you pause and wonder,

"How does this actually work?"

Data Centers, Servers & the Cloud

Most systems don't run on your phone.

They run in large buildings filled with thousands of computers working together.

When you search, stream, scroll, or ask a question, your request travels there and back in less than a second.

That network of computers is what people call "the cloud."

How AI "Learns"

AI doesn't understand things the way people do.

It doesn't know what something means.

It learns patterns, not people.

Instead, it learns patterns from lots of examples and predicts what comes next.

That's why it can give fast answers.

It's also why it can sound confident even when it's missing important context.

Recommendation Systems

These systems help decide what appears next in your feed, search results, videos, or online suggestions.

They watch what you click, watch, like, share, or spend time looking at.

Then they try to predict what you might want next.

In many cases, they're simply showing more of what they think will keep you interested.

Feedback Loops

Your actions teach the system.

The system shows you more of the same.

You interact more.

The cycle continues.

Over time, that cycle can influence what you notice, what you engage with, and sometimes even what you believe.

Deep fakes & Trust

Modern tools can create videos, voices, images, and messages that look or sound completely real.

Some are creative and harmless.

Others are designed to mislead.

That's why seeing something is no longer enough to know it's true.

Verification and questions matter more than ever.

The digital world can feel invisible.

But every system is built by people.

People decide what data to collect, what to optimize for, and what to ignore.

Understanding that doesn't answer every question.

It just helps you ask better ones.

Systems That Can Act (Agentic AI)

Some newer AI tools go beyond answering questions. They can plan steps, use other tools, and take actions on their own to complete tasks.

For example, they might search the web, write emails, or book appointments based on a goal you give them.

This is exciting because it can save time and help with complex work.

But it also means we need to be extra careful. When a system can act by itself, understanding what it's doing - and why - becomes even more important.

Where Everything Actually Happens

Most of the systems in this book don't live on your phone. They run in huge buildings full of computers - called data centers, that work together. When you search, scroll, or ask a question, your phone just sends the request. The real work happens somewhere far away, in what people call "the cloud." Your device is only the window. The thinking happens elsewhere.

How AI Learns

AI doesn't understand things the way people do. It learns by looking at thousands or millions of examples and finding patterns.

It gets good at predicting what comes next.. That's why it can finish sentences, recommend videos, or answer questions so quickly. But it also explains why it can sometimes sound completely confident... even when it's wrong or missing the full picture. AI doesn't truly "know." It predicts.

Why Feeds Feel So Personal

The videos, posts, and suggestions you see aren't random. The system watches what you click, how long you watch, what you skip, and slowly learns what keeps your attention. The more you engage, the better it gets at showing you more of the same.

The system isn't trying to understand you.

It's trying to predict what you'll do next.

Small Things That Matter

Data doesn't only come from what you type. It comes from what you click, how long you pause, where you go, and even what your phone or watch quietly measures. Over time, those tiny signals become patterns. That's how a system can catch possible fraud on your card or adjust an insurance premium - it's constantly comparing what's normal for you to what's happening right now.

Loops That Keep Going

Some systems don't just watch what you do. They influence it. They show you something you're likely to like, you spend more time on it, and the system learns to show you even more of it. This cycle can keep going without you noticing.

When Things Look Real (But Aren't)

With today's tools, it's getting easier to create videos, voices, and images that look and sound completely real. That's why simply seeing something is no longer enough. Trust now needs questions, not just eyes.

The Trade-Off

The more these systems know about you, the more helpful they can be. But that same information also gives them more power to shape what you see and do. There isn't always one right answer. Awareness is what helps you decide where to draw the line.

The Real-World Side

Even though everything feels digital, these systems still need real things - electricity, hardware, and big buildings.

As more people use them, the demand on those resources grows too.

The digital world can feel invisible. But it still depends on very real things.

HOW EVERYTHING CONNECTS

Your actions → Systems learn patterns → Predictions → Influence → New actions

Most of the time, this happens quietly.

In feeds.

In recommendations.

In rankings.

In decisions people barely notice.

But the more aware you are of the cycle, the more intentional your choices can become.

Because once you see the pattern, you start noticing it everywhere.

And once you notice it, you can decide for yourself what happens next.

FINAL THOUGHT

Maya started with a simple question. Why?

Why did one application get accepted while hers didn't?

Why did some decisions feel fair and others completely invisible?

Why did it sometimes feel like the systems knew her, but didn't really see her?

She never found perfect answers. And maybe that was okay.

What changed was simpler. She started noticing.

She noticed the systems all around her - the ones that shaped what she saw, what she could borrow, who got a chance, and what felt normal. Some helped. Some hurt. Most were more complicated than they first seemed.

But they weren't magic. They were built by people. They carried old stories. And they could be questioned.

The goal was never to fear them. It was to see them.

Because once you notice something, it's hard to stop seeing it.

In a world that moves fast and answers quickly, maybe the most important thing isn't having all the right answers. It's knowing when to pause. When to ask one more question. When to look a little closer and decide for yourself.

Curiosity still matters. Kindness still matters. The messy, human part - the part that dreams, doubts, and cares - still matters most.

The systems will keep changing. New tools will come. New challenges will follow.

But Maya learned she didn't have to let them decide everything. A score could influence a path. A system could shape a decision.

Even as new systems appeared - ones that could plan, act, and work toward goals on their own, Maya understood something important. No matter how smart the tools became, the power to choose still belonged to her.

And that felt like it was enough to begin.

DISCUSSION QUESTIONS

If you could ask any system one honest question and it had to answer truthfully - what would you ask?

Would you ask about a recommendation feed, a credit score, a scholarship decision, an AI tool, or something else?

1. When a system makes a decision without explanation

Maya's rejection didn't come with a clear reason. How would that feel to you? Would it change how much you trust the outcome?

2. Patterns vs your personal story

Systems rely on patterns from many people. Can you think of a situation where your personal story might not fit the pattern? What parts of a person do you think systems often miss?

3. What data misses

The system only sees what is measured. What are some important things about people that systems may never fully capture? Should important decisions rely only on measurable information?

4. Small actions, bigger outcomes

The book shows how small signals can influence bigger decisions over time. What are some everyday actions that might matter more than we realize? Why do repeated patterns become powerful?

5. When predictions become decisions

At what point do you think a prediction starts feeling like a decision? Who should question that? Who should be responsible for explaining it?

6. **Trust and confidence**

Have you ever trusted something because it sounded confident, urgent, or official? What made it believable? How do you decide what to trust online now?

7. **Social media and perception**

Aarohi's online presence looked complete - but wasn't. Why do you think people feel pressure to appear "fine" online? Have you ever seen something online that changed how you felt about yourself or others?

8. **Influence on behavior**

Systems don't just respond. They influence. Have you noticed recommendations, notifications, or trends shaping your attention or choices? When does something helpful become difficult to ignore?

9. **Speed vs understanding**

AI tools can provide answers instantly. Do faster answers always lead to better understanding? What parts of learning still matter even when information becomes easy to generate?

10. **What systems optimize**

For Many systems are designed to maximize something—attention, engagement, speed, efficiency, or profit. How do you think that affects what people see, feel, or believe? What happens when keeping attention becomes more important than helping people?

11. **Confidence online**

The book shows how confident systems, videos, and online opinions can sound - even when they may be incomplete or wrong. Why do you think confident information spreads so quickly online? How can people learn to pause before automatically trusting something?

12. The feedback loop

Think about this cycle: Your actions → system learns → predictions → influence → new actions. Where have you seen this happen in real life? Do people shape systems… or do systems slowly shape people too?

13. Human decisions

The book suggests that systems can assist decisions - but shouldn't fully replace people. What kinds of decisions should always involve humans, even if AI becomes very accurate? Why?

14. Noticing differently

After reading this book, is there anything you might pay more attention to now? Online, in school, in recommendations, or in everyday decisions?

15. Looking ahead

As AI systems become more powerful, what do you think people will need to protect most? Trust? Privacy? Creativity? Independence? Something else? Why?

16. Protecting family from scams

Maya helped her grandma avoid a phishing attempt and a deepfake voice call. Have you or someone you know ever received a suspicious message or call? What small steps can families take together to stay safer?

17. Privacy, compound interest, and the cloud

Small daily choices can compound over time—both in money and in the data we leave behind. How do you feel about trading convenience and privacy for better personalization or helpful recommendations? Where would you draw the line?

18. AI, jobs, and creativity

The story explores how AI can copy, predict, and create fast—but may not replace human dreams or messy ideas. How do you think AI will change jobs and creativity in the future? What parts of being human do you think will always matter most?

19. Systems That Act on Their Own

The story shows AI tools that can do more than answer questions - they can take actions and make decisions. How do you think this changes the way we should use them? What rules or questions should we have before letting a system spend money, send messages, make decisions, or act on our behalf?

Imagine Maya is five years older.

What advice do you think she would give her younger self at the beginning of this story?

You don't need to answer all of these. (Or any of them.) Sometimes, just noticing the question is where thinking begins.

DISCUSSION GUIDE

You don't need to answer every question. Some of the best thinking starts with just noticing one.

1. Fairness and Decisions Maya's scholarship rejection came without explanation. How would that feel? What makes a decision feel fair to you?

2. Patterns vs. People Systems see patterns, not full stories. Can you think of a time when your own story didn't fit the expected pattern? What important things about people do systems often miss?

3. What Data Remembers Small daily choices—clicks, scrolls, payments—can shape future outcomes. What habits are you building today that might matter years from now?

4. Trust and Confidence The book shows how confident answers, posts, and messages can feel convincing. Have you ever trusted something because it sounded sure? How do you decide what to trust now?

5. Social Media and Reality Aarohi's online life looked perfect, but her real moments were different. Why do people feel pressure to show only polished versions of themselves? What gets left out?

6. Attention and Influence Feeds and games are designed to keep us engaged. Where have you noticed systems quietly shaping what you watch, buy, or do?

7. AI and Thinking Tools can give fast answers, but Maya learns they shouldn't replace our own thinking. When is it helpful to use AI? When is it better to figure things out yourself?

8. Protecting Others Maya helped her grandma avoid a scam. What small steps can families take together to stay safer online?

9. Looking Ahead As AI becomes more powerful, what do you think we should protect most - privacy, creativity, jobs, fairness, or something else? Why?

10. One Small Change After reading Maya's story, is there one thing you might notice or do differently when using apps, social media, or AI tools?

11. If an AI can take actions instead of just answering questions, what responsibilities come with using it?

CONTINUE THE CONVERSATION

- Share your thoughts about Maya's journey
- What invisible systems have you noticed?
- Discussion resources for parents and educators

AUTHOR'S NOTE

This story isn't really about technology.

It's about the quiet systems we don't always see - the ones shaping decisions in school, money, social media, and everyday life.

The idea for this book came from conversations at home. My kids kept asking honest questions - about results that didn't make sense, about things that showed up on their screens, and about decisions that felt… already made. Those conversations reminded me how important it is for young people to understand the world they're growing up in.

This book was completed during the year of America's 250th anniversary. While technology continues to change how we learn, work, and make decisions, the questions Maya explores are much older: How do we decide what is fair? Who gets opportunities? How do we balance innovation with responsibility? And how do we remain thoughtful and human in a rapidly changing world?

Through Maya's eyes, you saw how these systems work -where they helped, where they fall short, and where they miss parts of the human story. These systems don't fully know us. They learn patterns from the past and use them to predict the future. Sometimes those predictions are helpful. Sometimes they're not.

This is not a textbook. It's a story meant to help you notice what often goes unseen. Because once you notice, you begin to question. And once you question, you begin to choose.

AI is powerful, but it is also limited. Like any tool, it isn't good or bad on its own. What matters is how we understand and use it. I hope this book helps young readers, parents, and teachers explore these questions together.

In a world moving faster every day, the ability to think critically, ask better questions, and remember our humanity may be one of the most important skills we can have.

Thank you for reading. I hope this story leaves you curious, thoughtful, and a little more confident in navigating the world around you.

Thank you for reading.

If this story made you pause...
If it made you question...
If it made you look a little closer at the invisible systems around, you...

Then Maya's journey has done its job.

Stay Curious.

-Subhasit Ratnam
Author website: www.subhasit.com

ABOUT THE AUTHOR

Subhasit Ratnam (Agrawal) currently works in technology and risk at a large financial institution, with years of experience in cloud, data, models, and AI systems. He holds a master's degree in computer engineering and Artificial Intelligence and has spent much of his career working with the kinds of systems Maya encounters throughout the story.

But this book didn't start at work.

It started at home—through everyday conversations with his kids, who kept asking honest, curious questions about how the world really works.

The Invisible Score grew out of those moments. It was written to turn complicated ideas about data, AI, and digital systems into a story that feels real - not a lecture.

He hopes this book encourages readers to pause, question, and think more clearly about the systems around them.

Outside of his professional life, Subhasit is passionate about giving back to the community through volunteering, pro bono work, and supporting efforts that help young people, parents, and educators better understand the digital world they're growing up in.

This is his first young adult novel.

ACKNOWLEDGMENTS

This book started with questions.

Thank you to my kids (Shlok & Prisha)—for the curiosity that inspired it, and for the questions that kept going.

Thank you to my wife (Payal), parents and family for the time, patience, and support along the way.

I'm grateful to my friends, colleagues, mentors, educators, and many others who have shared their knowledge, experiences, and ideas with me over the years, helping shape my understanding of technology, risk, and decision-making.

And to every reader - thank you for taking the time to notice.

Curiosity is where every important question begins.

www.ingramcontent.com/pod-product-compliance
Lightning Source LLC
LaVergne TN
LVHW022225200926
848481LV00006B/2125